AQUEOUS TWO-PHASE SYSTEMS

PROPERTIES, FUNCTIONS AND ADVANTAGES

ENVIRONMENTAL SCIENCE, ENGINEERING AND TECHNOLOGY

AQUEOUS TWO-PHASE SYSTEMS

PROPERTIES, FUNCTIONS AND ADVANTAGES

VANGELIS ANATOLIJS XANTHOPOULOS

EDITOR

NOTICE TO THE READER

Library of Congress Cataloging-in-Publication Data

ISBN: 978-1-53614-241-9

Published by Nova Science Publishers, Inc. † New York

CONTENTS

PREFACE

Aqueous Two-Phase Systems (ATPS) have been used since their discovery for the isolation of a large number of different biological materials, such as animal and plant cells, microorganisms, viruses, cellular organelles such as chloroplasts, mitochondria, membrane vesicles, and also in the purification of proteins and nucleic acids. In *Aqueous Two-Phase Systems: Properties, Functions and Advantages*, the authors begin by discussing the benefits of these systems.

The following chapter presents the main forms of the purification of lipases from aqueous biphasic systems, addressing and discussing the fundamentals for the formation, use and optimization of these systems applied in the biotechnological context. The authors address the basis of aqueous-biphasic systems, aqueous two-phase systems for enzymatic purification, purification of lipases and combined uses of aqueous two-phase systems.

The most important experimental parameters that affect the efficiency of metal ions extraction in aqueous PEG-based two-phase systems are examined to provide a starting point in the design of a suitable system for the extraction of metal ions. Compared to the conventional solvent extraction, which uses toxic, flammable and volatile organic solvents and can be quite expensive, the extraction in aqueous PEG-based two-phase systems is considered a more environmental friendly and economically viable method.

Also in this compilation, an overview about application of ATPSs in microstructured devices is provided. Microstructured devices offer potential benefits due to well-defined high specific interfacial areas available for heat and mass transfer. These areas increase transfer rate and enhance yield, selectivity and process control. The application of ATPSs in Micro Total Analysis Systems (μ-TAS) is described as well.

Chapter 1 - The Aqueous Two-Phase Systems (ATPS) have been used since their discovery for the isolation of a large number of different biological materials, such as animal and plant cells, microorganisms, viruses, cellular organelles such as chloroplasts, mitochondria, membrane vesicles, and also in the purification of proteins and nucleic acids. ATPS are produced by the combination of two chemically dissimilar water-soluble polymers, or a water-soluble polymer (polyethylene glycol) and a salt (potassium phosphate or citrate), which are added to water, resulting in the formation of two immiscible phases. Due to the high water content, aqueous systems provide a safe separation and high purification of biomolecules. Compared to chromatographic methods, ATPS are simpler and faster for the separation of molecules. The distribution balance takes place in a short time, they present low cost and offer the possibility of being applied on a large scale. In addition, they allow the design of different extraction stages, for which several extractive steps are carried out to increase the purity of the protein to be purified. Another positive aspect of the ATPS lies in the fact that both immiscible phases have a high water content (between 85 - 90%), so they have very low interfacial tension and the proteins do not undergo denaturing processes like the one that occurs in an air-liquid interface. At the same time, by adding salts or solutes with buffer capacity, both the pH and the ionic strength can be regulated. These characteristics provide the adequate environment that allows to preserve the biological activity of extremely labile materials, that is, they provide high stability to the macromolecules in both phases. This method of purification stands out among the main methods of precipitation, which can be used as a single purification stage, depending on the degree of final purity that is intended to be obtained. Its field of application has gained great acceptance in the scientific community due to its versatility, its functionality in the separation of biomolecules and its possibility of scaling up.

Chapter 2 - Lipases are important biocatalysts belonging to the group of hydrolases, presenting the ability to catalyze total or partial hydrolysis of triacylglycerol to diacylglycerol, monoacylglycerol, glycerol and even free fatty acids. In general, lipases are applied in a variety of processes due to their affinity with a wide range of substrates, as well as their thermal, pH and organic solvents stability. In the production of enzymes using biotechnological processes, such as lipases, the crude enzymatic extract obtained is composed of an aqueous mixture of cells, extra and intracellular products, substrates and unconverted components. After the fermentation or extraction, depending on its application, enzyme purification operations are required, which should be included in the final cost of the product. Processes that allow the recovery and purification of enzymes efficiently and at low cost are extremely desirable and represent a research area of great interest. Aqueous biphasic systems are based on a liquid-liquid fractionation technique and are widely studied for the recovery and purification of enzymes, such as lipases. This technique involves the construction of extraction systems that can be formed by: (i) two polymers; (ii) a polymer and a salt; (iii), an ionic liquid and a salt or (iv) an alcohol having a low molar mass and a salt mixed over a concentration limit, thereby forming two immiscible phases. This method is widely used in protein purification because it minimizes the denaturation and/or loss of biological activity due to the high water content and low interfacial tension of the systems, which protect the proteins. In addition, two-phase aqueous systems are applicable in real scale and allow recycling of the reagents used in the process. Its high water content implies in high biocompatibility and low interfacial tension, minimizing the degradation of biomolecules. In this way, the present chapter will present the main forms of purification of lipases from aqueous biphasic systems, addressing and discussing the fundamentals for the formation, use and optimization of these systems applied in the biotechnological context. For this the following items will be addressed: i) Introduction; ii) Basis of aqueous-biphasic systems; iii) Aqueous-two-phase systems for enzymatic purification; iv) Purification of lipases; v) Prospects and combined uses of aqueous two-phase systems.

Chapter 3 - Compared to the conventional solvent extraction, which use toxic, flammable and volatile organic solvents, requiring large samples and large volumes of extractants and organic solvents, and can be quite expensive, the extraction in aqueous PEG-based two-phase systems is considered a more environmental friendly and economically viable method. This is because aqueous two-phase systems do not require the use of organic solvents, is efficient (the extraction percent can exceed 99% in well-defined experimental condition, and has low cost (most of the phase-forming components are commercially available and no so expensive). Generally, the aqueous two-phase systems are obtained by mixing an aqueous solution of certain water-soluble organic polymer (most frequently used is polyethylene glycol, PEG) with certain inorganic salts, in specific concentration. The obtained extraction systems are composed by two aqueous immiscible phases, the top one – rich in PEG, with the same role as organic phases from traditional extraction systems, and a bottom phase – salt-enriched. The extraction of metal ions in such aqueous two-phase systems depends on (i) the formed aqueous two-phase system characteristics and (ii) the properties of formed metallic species in extraction system. If the optimum characteristics of the aqueous two-phase system can be obtained via suitable selection of phase forming components, the second condition required the utilization of some extracting agents which will provide the selectivity of extraction process. In this chapter, are discussed the most important experimental parameters that affect the efficiency of metal ions extraction in aqueous PEG-based two-phase systems, to provide a starting point in the design of a suitable system for the extraction of metal ions.

Chapter 4 - Aqueous two-phase systems (ATPSs) are formed when two incompatible polymers or certain polymers and lyotropic salts are mixed. They have been recognized as superior and versatile, preparative and analytical tools for the downstream processing of biomolecules with wide application in extractions, separations, purifications and enrichments of proteins, polyphenols, viruses, enzymes and other biomolecules both in research and industry. An overview about application of ATPSs in microstructured devices is given in this work. Microstructured devices offer potential benefits due to well-defined high specific interfacial areas available for heat and mass transfer. High specific interfacial area increases transfer rate and enhances yield, selectivity and process control. The flow patterns formations and stability, mass transfer and samples partition coefficients using ATPSs in microstructured devices are discussed in this work. Also, the application of ATPSs in Micro Total Analysis Systems (μ-TAS) is described.

In: Aqueous Two-Phase Systems
Editor: V. Anatolijs Xanthopoulos

ISBN: 978-1-53614-241-9
© 2018 Nova Science Publishers, Inc.

Chapter 1

THE COMPONENTS, FUNCTIONS AND ADVANTAGES OF AQUEOUS TWO-PHASE SYSTEMS

Maryen Alberto Vazquez[*]
Department of Physiology and Biochemistry of the Institute of
Animal Science, Mayabeque, Cuba

ABSTRACT

The Aqueous Two-Phase Systems (ATPS) have been used since their discovery for the isolation of a large number of different biological materials, such as animal and plant cells, microorganisms, viruses, cellular organelles such as chloroplasts, mitochondria, membrane vesicles, and also in the purification of proteins and nucleic acids. ATPS are produced by the combination of two chemically dissimilar water-soluble polymers, or a water-soluble polymer (polyethylene glycol) and a salt (potassium phosphate or citrate), which are added to water, resulting in the formation of two immiscible phases. Due to the high water content, aqueous systems provide a safe separation and high purification of biomolecules. Compared to chromatographic methods, ATPS are simpler and faster for the separation of molecules. The distribution balance takes

[*] Corresponding Author Email: mvazquez@ica.co.cu.

place in a short time, they present low cost and offer the possibility of being applied on a large scale. In addition, they allow the design of different extraction stages, for which several extractive steps are carried out to increase the purity of the protein to be purified. Another positive aspect of the ATPS lies in the fact that both immiscible phases have a high water content (between 85 - 90%), so they have very low interfacial tension and the proteins do not undergo denaturing processes like the one that occurs in an air-liquid interface. At the same time, by adding salts or solutes with buffer capacity, both the pH and the ionic strength can be regulated. These characteristics provide the adequate environment that allows to preserve the biological activity of extremely labile materials, that is, they provide high stability to the macromolecules in both phases. This method of purification stands out among the main methods of precipitation, which can be used as a single purification stage, depending on the degree of final purity that is intended to be obtained. Its field of application has gained great acceptance in the scientific community due to its versatility, its functionality in the separation of biomolecules and its possibility of scaling up.

Keywords: water-soluble polymers, macromolecules, purification

INTRODUCTION

Macromolecule Purification, Generalities

The industrial use of different macromolecules has increased in recent years, and it is therefore necessary to develop new methods of isolation and purification with a considerable purity, low cost and that can be applied on a large scale. The conventional methods of purification of macromolecules consist of the following fundamental steps:

1. Rupture of cells and removal of insoluble material by centrifugation and filtration.
2. Precipitation fractionated with organic solvents or salts.
3. Purification of the molecule of interest by ion exchange chromatography, affinity, etc.
4. Concentration of the biomolecule solution by dialysis and ultracentrifugation.

These steps consume too much time and cause adverse conditions that can denature the macromolecule and significantly reduce the performance of the process. The reagents used generally cannot be recycled and must be discarded to the environment with the consequent impact on it. The process of isolation and purification of a protein is a major problem due to the complexity of the protein mixtures, the need to maintain biological activity, and the high cost-performance ratio that must also allow the recycling of the reagents used in the proteins successive steps.

Diamond and Hsu (1989) postulated that between 50 - 90% of the total cost of production of a biological product is determined by the purification process. Traditional protein purification procedures include several stages of precipitation. For mixtures with low amount of contaminants and high concentration of the protein of interest it can be an effective method to concentrate, and in some cases the precipitation, can act as a single purification stage, depending on the degree of final purity to be obtained (Pesso and Vahan, 2005).

Precipitation is a widely used operation, not only on a laboratory scale but also on an industrial scale, for the purification of proteins of microbial, animal or vegetable origin (Kim et al., 2000, Cooper et al., 2005, Boeris et al., 2008). It has been used for decades, for example, for the precipitation of milk casein by the action of dilute mineral acids.

In this group, a novel method is being investigated in the last years and is based on the extraction of biomaterial using Aqueous Two-Phase Systems (ATPS). This method is considered very useful for the purification process of macromolecules (Oliveira et al., 2002, Antov et al., 2005, Klomklao et al., 2005, Su and Chiang, 2006, Du et al., 2007).

Aqueous Two-Phase Systems

The ATPS have been used since its discovery for the isolation of a large number of different biological materials such as animal and plant cells, microorganisms, viruses, chloroplasts, mitochondria, membrane vesicles, proteins, nucleic acids, etc. (Zaslavsky, 1995).

ATPS are produced by the combination of two chemically different water-soluble polymers, or a water-soluble polymer (polyethylene glycol) and salt (potassium phosphate or citrate), added to water, resulting in the formation of two rich immiscible phases.

Due to the higher water content, ATPS have several advantages compared to common-use separation and purification techniques because they provide a safe separation and technical purification of biomolecules (Kim et al., 2001; Seyrek et al. 2003, Cooper et al. 2005; Cooper et al. 2006; Basak et al. 2007).

Compared to traditional methods, ATPS turn out to be simpler and faster for separation, since the distribution equilibrium takes place in a short time, they present low cost and the possibility of being applied on a large scale. They also allow the design of different extraction stages: several extractive steps can be carried out in order to increase the purity of the protein. It is also possible to control the volumes of the phases to concentrate the protein of interest at the same time that it is being purify. A continuous extraction could be made, taking advantage of the removal of one of the phases (Brooks et al., 1985; Walter and Johanson, 1986).

The successive extractive steps consist of removing the phase containing the impurities of the first biphasic system and placing a clean phase in its place so that it is balanced with the phase containing the biological product of interest and thus reestablishing a second distribution equilibrium. In the case of applying successive extractive steps, the loss of yield on the increase in purity must be analyzed in order to reach a suitable commitment relationship (Palomares, 2004).

Another positive aspect of the ATPS lies in the fact that both immiscible phases have a high water content (within the range 85 - 90%) so they have very low interfacial tension and the proteins do not undergo denaturation processes such as the one that occurs in an air-liquid interface. At the same time, by adding salts or solutes with buffer capacity, both the pH and the ionic strength can be regulated. These characteristics provide the adequate environment that allows to preserve the biological activity of extremely labile materials, that is, they provide high stability to the macromolecules in both phases. Finally, it is important to consider that one of the phases can be achieved as a means of incubation of a bioreactor (microorganism in which the protein of interest is expressed) and the other as a receptor of the protein once it is excreted into the medium.

ATPS Types

In general, ATPS types can be joined in two large groups.

Preformed

They are formed when: Two different water-soluble polymers solutions mixing above a certain critical concentration of the polymers. In the ATPS formed, each phase is enrich in one of the polymers. Besides, when a solution of water-soluble polymers and an inorganic salt mixing above a certain critical concentration value. One of the phases is rich in polymer and the other in salt. Within this group polyethylene glycol (PEG) -phosphate (Pi) systems are very attractive due to the low cost and proteins large-scale separation.

Induced by Temperature

When solutions of certain water-soluble polymers are heated above a certain critical temperature called "cloud point" or phase separation temperature (TSF), two phases are separated, one rich in the polymer and another phase formed by a very dilute aqueous solution of said polymer. The value of the phase separation temperature depends on the structural characteristics of the water-soluble polymers.

The most used are copolymers of ethylene oxide and propylene oxide (EO-PO) whose phase separation temperatures are in the range 5 - 60°C, being useful for the separation of biological molecules. PEG can also be considered a thermo-reactive polymer, but its phase separation temperature is too high (above 100°C) to be used in the separation of labile biomolecules.

Recently, the isolation and purification of certain biomolecules has been carried out successfully thanks to the combination of a two-phase system preformed with one induced by temperature. In the first stage, the material containing the molecule of interest is distributed in a preformed system with a thermo-sensitive polymer and another water-soluble polymers. The conditions of the medium (pH, ionic strength, presence of salts) are regulated so that the molecule of interest is preferably distributed to the phase rich in the thermo-sensitive Polymer.

Then this phase is removed and heated above the TSF. This leads to a new two-phase system (thermo-induced) where one of the phases is practically an aqueous solution of the protein of interest and the other is a concentrated solution of the thermo-sensitive polymer, which is thus recycled and can be used in a new extractive step.

Flexible Chain Polymers

The flexible chain polymers (FCP) have simple bonds between the carbon atoms that form the hydrocarbon chain, so that the alkyl residues (-CH2-CH2-) can rotate freely. This gives the molecule the quality of being totally flexible and therefore of being able to acquire any conformation at random. The FCP that are used in two-phase aqueous systems are those that are soluble in water and that generally do not have an electrical charge.

They can be classified according to their source in:

(a) Natural: they come from the partial hydrolysis of natural polymers insoluble in water, such as starch and cellulose. Of these, water-soluble derivatives are obtained, such as: hydroxy propyl starch, alkyl celluloses and 1-4 maltodextrin. We can also mention within this group the derivatives of the enzymatic polymerization of carbohydrate hydrates: dextran and ficoll. All are biodegradable, which gives them great interest because their use and subsequent disposal does not cause environmental impact.

(b) From industrial synthesis: polyethylene glycol, polypropylene glycol, polyacrylamide, polycaprolactan, copolymers of ethylene and propylene oxides. They are generally not biodegradable, and in some cases they can be toxic (Albertsson et al., 1987).

Binodial Diagrams

The formation of an aqueous two-phase system occurs when mixing solutions of two polymers or a polymer and a salt, only when they are present in a certain concentration range. The concentrations of polymers (or polymers and salts) in which phase separation occurs, can be represented in a "phase diagram" where the vertical axis represents the polymer from which the upper phase is enriched, and the horizontal axis the salt (or the polymer) that enriches the lower phase.

The so-called "binodial curve" separates two zones: the area between the axes of coordinates and the curve, which corresponds to single-phase systems, and the mixtures whose concentration in polymers (or FCP and salt) correspond to points located above the curve, which will lead to the formation of two aqueous phases, Figure 1.

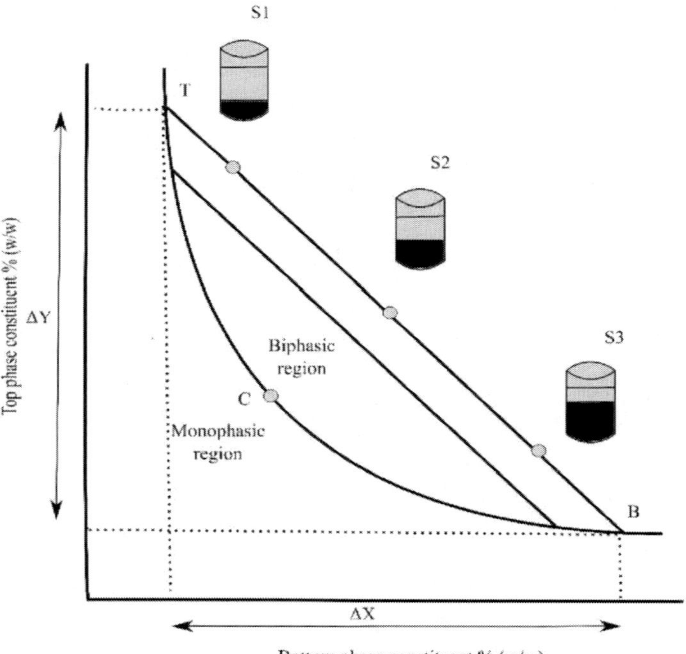

Figure 1. Binodal curve, TCB = Binodal curve, C = critical point, TB = Tie line, T = composition of the top phase, B = composition of the bottom phase, and X, Y and Z = total composition of ATPS.

Factors that are Responsible for the Distribution of a Protein in an ATPS

The factors that affect the distribution of a biomolecule in an ATPS are diverse. There is a complex interaction of the substance to be separated and the molecules that surround it (solvent molecules, FCP, etc.), hydrogen bonds, electrostatic, hydrophobic interactions, Van der Waals forces can be formed, as well as steric effects.

It is expected, therefore, that the molecular weight and chemical properties of the polymers in each phase have an influence on the distribution of these biomolecules. Other factors that also influence the distribution are the size and superficial properties such as hydrophobicity, surface charge of the biomolecules submitted to the distribution, as well as their conformation.

Composition of the Phases

Near the critical point the compositions of the equilibrium phases are similar, and the partition coefficient of a given biomolecule approaches unity. On the contrary, far from the critical point, each phase is significantly enriched in one of the two polymers and deficient in the other. Due to these differences in the composition of the phases, the distribution of the molecule of interest is unequal, and the interfacial tension between the two phases is increased.

Molecular Weight of Polymers

It has been observed that the variation in molecular weight (MW) of the polymers influences the distribution of the biomaterial for two reasons: modifying the phase diagram and changing the number of protein-polymer interactions.

It has been observed that an increase in the molecular weight of the more hydrophobic PCF decreases the tendency of the protein to partition in the phase that is enriched with said polymer, increasing the tendency of the protein to be distributed in the phase containing the most hydrophilic polymer. While the increase in molecular weight of the most hydrophilic polymer causes the protein to be distributed to the rich phase in that same polymer.

These results have an easy explanation based on the preferential interaction. In the case of polyethylene glycol, its molecular weight has a marked influence on the distribution of proteins. It has been reported that high molecular weight polyethylene glycol has a stabilizing effect on proteins with respect to the low molecular weight polymer. This fact argues in favor of steric exclusion when determining the interaction of polyethylene glycol with proteins.

Effect of the Hydrophobicity of the Protein

It is known that proteins present in their native conformation most of the non-polar residues inside it, however, there is a significant amount of hydrophobic residues exposed to the solvent. The difference in hydrophobicity affects the distribution since it is related to the surface properties and has been used as a measure of the relative surface hydrophobicity of different proteins. In this sense, for example, denatured proteins are distributed differently, not only because their exposed area is greater but also because it is much more hydrophobic. Hydrophobicity.

Summarizing, there are two main effects: phase hydrophobicity effect and salting out effect, are involved in hydrophobic interactions. In polymer – salt systems, hydrophobicity may be manipulated by varying MW of polymer and by adding a salt (e.g., NaCl). The low NaCl concentrations (< 1 M) do not affect ATPS however, high salt concentrations (> 1 M) changes the phase diagram [Andrews and Asenjo, 2010). The addition of salt in ATPSs has a significant effect on the partition coefficient. These salts contain ions of different hydrophobicities and the hydrophobic ions force the partitioning of counter ions to phase with higher hydrophobicity and vice versa. The salting-out effect moves the biomolecule from salt-rich phase to polymer-rich phase (Raja et al. 2011).

                    Maryen Alberto Vazquez

Net Electric Charge of the Macromolecule, pH and Ionic Composition of the ATPS

The chemical potential of a protein in a given phase is a function of its activity, its surface energy and surface area, its net charge and the difference in electrostatic potential between both phases.

The pH of ATPS may alter the charge and surface properties of solute which affects the partitioning of biomolecule. The net charge of the protein turns negative in case of higher pH than the isoelectric point (pI) and positive if lesser than pI. If the pH is equal to pI, net charge will be zero (Raja et al. 2011). It has been reported that the partitioning of negatively charged biomolecule in a higher pH system increases the partition coefficient and target biomolecule prefers top phase. Higher pH values than pI of protein induce an affinity towards PEG-rich phase because of the positive dipole moment (Andrews et al. 2005).

Raising the chemical potential in phases 1 and 2 for the protein, defining the protein partition coefficient Kp and defining the K°p as the partition coefficient at zero electrostatic potential, or when the protein is at its isoelectric point so that the net charge is zero. The pH change in addition to modifying the net charge of the protein can also induce conformational changes in the structure of the protein that affect its distribution.

At extreme pH, the proteins can be denatured and acquire different distribution properties to those of the native state, since their surface is more hydrophobic and they have a significantly higher exposed area.

Effect of Temperature

Temperature greatly affects the composition of two phases in an ATPS, hence, the phase diagram. The changes in temperature also affect partition through viscosity and density. Therefore, it is always recommended by the researchers to have a strict control of temperatures in ATPS related experiments. In general, phase separation is obtained at lower temperature in a polymer – polymer ATPS with lower concentrations of polymer, however, an opposite effect is seen in polymer – salt system (Walter, 1994).

Partitioning behavior of biomolecule and phase separation rate is also influenced by the physico-chemical properties (i.e., density, viscosity and interfacial tension) of ATPS. Measurement of such properties have been explained by Albertsson (1986), Zaslavsky (1994) and Hatti-Kaul (2000)

The temperature affects the distribution process based on:

- A change in the composition of the flexible polymers in both phases.
- Inducing conformational changes in the structure of the protein.

In general, the processes of protein distribution are endothermic. Therefore, in polyethylene glycol dextran systems, the increase in temperature favors the distribution of the protein in the phase rich in polyethylene glycol.

Functions and Advantages

Enzymes and Proteins Separations

Several research articles regarding ATPS for protein partitioning have been published. Fisher et al., reported nanoparticle mediated protein separation in micellar systems with high recovery yields (Fischer et al. 2012, 2013). IL-based ATPSs are being used in current scenario [77]. However, most recently, Li et al., showed the extraction of protein using deep eutectic solvent (DES) – ATPS. They selected Betaine-urea for extraction and studied the influence of pH, temperature, salt concentration, separation time, amount of protein and the mass of DES. More importantly, the conformation of protein was not changed and the extraction efficiency was 99.82%. All the results suggested betaine-based DES – ATPS a potential new system for protein separation (Li et al. 2016).

In the case of enzymes, conventional liquid-liquid extraction techniques are generally not applicable for the purification due to the irreversible loss of enzymatic activity (Walter, 1994). Other constraints of these traditional purification techniques are laborious procedures and low yields (Saxena et al. 2003, Ramakrishnan et al. 2016). Ultrafiltration and hydrophobic interaction chromatography are also multi-step, costly and low-yielding purification techniques. Affinity and ion exchange chromatography are the two most frequently used chromatography techniques for the purification of recombinant enzymes. But a sample pre-treatment step is required to facilitate the capture of the target protein. Ion exchange chromatography requires a desalting step while using LB media, which is the commonly used media and contain salt and yeast extract. Metal affinity chromatography is also impossible as yeast extract is known to contain cysteine rich protein (Rahimpour et al. 2016). Thus, ATPS was developed to overcome such limitations (Li et al. 2014, Rajagopalu et al. 2016). It is also capable of combining different downstream processes into a single step (Rajagopalu et al. 2016, Madhusudhan et al. 2008).

Mainly biphasic systems have been used in the purification of enzyme such as proteases (Alberto et al. 2016) Laccases, which are found in plants, insects, bacteria and fungi. Fungal laccases have been purified and characterized extensively. They are successfully being used for pharmaceutical products (e.g., antibiotics, anticancer drugs, anesthetics and sedatives). Recently, in 2016, Rajagopalu et al., purified laccase from Hericium erinaceusi using PEG8000 – potassium phosphate ATPS and achieved a recovery 99% (Rajagopalu et al. 2016). Xylanases enzymes have been used in various applications, one of the major use is in the pulp and paper production as it can degrade the backbone of hemicellulose xylan (Rahimpour et al. 2016, Mamo et al. 2006). In 2016, Rahimpour et al., studied the two-stage purification of xyalanase using a 6% PEG 6000–20% phosphate system at first stage. Results showed about 78% of recovery after final separation and a purification factors of 6.7 (Rahimpour et al. 2016).

Packed columns are being used for enzyme purification because they require small space and are more efficient because of reduced axial mixing. A recovery of 94% of xylanase was achieved by Igarashi et al., by PEG 4000 – dipotassium phosphate ATPS (Igarashi et al. 2004). Elastase extraction from Bacillus sp. EL31410 was done by Xu and co-workers using an optimized ATPS of 23.1% PEG2000 – 11.7% KH2PO4/K2HPO4 with a predicted recovery of 89.5% (Xu et al. 2005).

ATPS are an economical and environmentally stable option for the purification also of others important proteins such as immunoglobulins (Muendges et al. 2015) and monoclonal antibodies (Yang et al. 2008).

Others Macromolecules

The development of molecular biology and gene therapy is also dependent on the efficient and cost-effective isolation of DNA, RNA and other nucleic acid based biomolecules, cells and organelles. Frerix and co-workers reported such low cost, polymer–salt (PEG – potassium phosphate) for the recovery of plasmid DNA (pDNA) in combination with ultrafiltration. ATPS was composed of 15% PEG and 20% potassium phosphate and strong partitioning of pDNA was observed at pH 7.0 (Frerix et al. 2005).

ATPS has been successfully employed for the recovery of virus, virus like particles (VLPs) or bionanoparticles however, has not been fully explored yet (Nestola et al. 2015). In addition, they have been used for isolation of low molecular weight biomolecules (e.g., phytochemicals and secondary metabolites) that are considered as high valued products due to their broad applications in food to pharmaceutical industry (Moo &Young, 2011).

Environmental Remediation

The applications of ATPS are not only limited to biotechnology, but also being extended to environmental remediation in various ways according to Hatti (2000, 2001).

As we know the ATPS is an economic and eco-friendly method, used for the removal of textile dyes, and large-scale removal of microorganism (Huddleston et al. 1999; Mageste et al. 2009). Other examples include the removal of metal ions (Rogers et al. 1998), food-coloring dyes (Huddleston et al. 1998; Raghavarao et al. 2003) aromatics from industrial and environmental settings (Willauer et al. 1999).

Other Applications

ATPS has been most commonly used for the separation of macromolecules; however, it can also be employed as an analytical tool. The analytical applications (e.g., understanding chemical properties and behavior of protein, etc.) of ATPS are possible because the distribution between phases depends on system variation instead of the distribution of small molecules (Hatti, 200). On the other hand, partitioning in ATPS is sensitive to surface properties and conformation of phase forming component and particulate material (Hatti, 2001). According to Grilo et al. (2016), ATPS is gaining interest as an analytical tool and 8% of total application articles published in 2013 (ISI-indexed) account for analytical applications.

ATPS has been used with devices (column contactors) for continuous processes and microfluidics devices for the fractionation of biomolecules (Toner et al. 2005, Soohoo et al. 2009). Moreover, ATPS can also be integrated with other separation techniques such as solvent sublation, which is termed as aqueous two-phase flotation (Show et al. 2015) ATPS has also been successfully integrated with centrifugal partition chromatography (CPC) for the separation of biomolecules (Ivetic et al. 2013).

CONCLUSION

ATPS is a simple, selective and low cost promising separation technique. Easy scalability of this technique makes it valid to be adopted by industries for downstream processing. This separation technology allows its use for a wide range of functions. Above all, it should be specified that in the case of its use for biological samples and in the purification of enzymes its friendly environment is vital for the stability of the macromolecules. In a general way, it is a method that has a high development potential besides its integration with other promising tools will also be a breakthrough for recovering high value products.

REFERENCES

Albertsson, PÅ. *Partition of Cell Particles and Macromolecules*. Hoboken: Wiley Intersciences, 1986.

Albertsson, PA; Cajarville, AD; Brooks, E. Partition of proteins in aqueous two phase systems and effect of molecular weight of the polymer. *Biochimica Biophysica Acta.*, 1987, 926, 89-93.

Andrews, BA; Asenjo, JA. Theoretical and experimental evaluation of hydrophobicity of proteins to predict their partitioning behavior in aqueous two phase systems: a review. *Sep Sci Technol.*, 2010, 45, 2165–70. doi: 10.1080/01496395.2010.507436.

Andrews, BA; Schmidt, AS; Asenjo, JA. Correlation for the partition behavior of proteins in aqueous two-phase systems: effect of surface hydrophobicity and charge. *Biotechnol Bioeng.*, 2005, 90, 380–90. doi: 10.1002/bit.20495.

Antov, MG; Pericin, DM; Dasic, MG. Aqueous two-phase partitioning of xylanase produced by solid-state cultivation of Polyporus squamosus. *Process Biochem.*, 2006, 41, 232–235.

Basak, A; Strand, SP; Tribet, C; Jaeger, W; Dubin, PL. Effect of polyelectrolyte structure on protein–polyelectrolyte coacervates: coacervates of bovine serum albumin with poly (diallyldimethyl-ammonium chloride) versus chitosan. *Biomacromolecules*, 2007, 8, 3568–77.

Boeris, V; Spelzini, D; Peleteiro, J; Picó, G; Romanini, D; Farruggia, B. Chymotrypsin–poly vinyl sulfonate interaction studied by dynamic light scattering and turbidimetric approaches. *Biochim Biophys A.*, 2008, 1780, 1032–7.

Brooks, W; Walter, H; Fisher, D. *Partitioning in Aqueous Two-Phase Sytems Theory, Methods, Uses, and Applications to Biotechnology.* Edited by Harry Walter, Donald E. Brooks and Derek Fisher. Academic Press, Orlando, Florida, USA., 1985, pp. 21-34.

Cooper, CL; Dubin, PL; Kayitmazer, AB; Turksen, S. Polyelectrolyte–protein complexes. *Curr Opin Colloid Interface Sci.*, 2005, 10, 52–78.

Cooper, CL; Gouldin, A; Kayitmazer, B; Ulrich, S; Stoll, S; Turksen, S. Effects of the polyelectrolyte chain stiffness, charge mobility, and charge sequences on binding to proteins and micelles. *Biomacromolecules.*, 2006, 7, 1025–1035.

Du, Z; Yu, Y; Wang, JH. Extraction of proteins from biological fluids by use of an ionic liquid/aqueous two-phase system. *Chem Eur J.*, 2007, 13(7), 2130–2137.

Fischer, I; Franzreb, M. Nanoparticle mediated protein separation in aqueous micellar two-phase systems. *Solvent Extr Ion Exc.*, 2012, 30, 1–16. doi: 10.1080/07366299.2011.581093.

Fischer, I; Hsu, CC; Gärtner, M; Müller, C; Overton, TW; Thomas, OR; Franzreb, M. Continuous protein purification using functionalized magnetic nanoparticles in aqueous micellar two-phase systems. *J Chromatogr A.*, 2013, 1305, 7–16. doi: 10.1016/j.chroma.2013.06.011.

Frerix, A; Müller, M; Kula, MR; Hubbuch, J. Scalable recovery of plasmid DNA based on aqueous two-phase separation. *Biotechnol Appl Biochem.*, 2005, 42, 57–66. doi: 10.1042/BA20040107.

Grilo, AL; Raquel Aires-Barros, M; Azevedo, AM. Partitioning in aqueous two-phase systems: fundamentals, applications and trends. *Sep Purif Rev.*, 2016, 45, 68–80. doi: 10.1080/15422119.2014.983128.

Hatti-Kaul, R. Aqueous two-phase systems. *Mol Biotechnol.*, 2001, 19, 269–77. doi: 10.1385/MB:19:3:269.

Hatti-Kaul, R. *Aqueous Two-Phase Systems: Methods and Protocols.* Berlin: Springer Science & Business Media, 2000.

Huddleston, JG; Ingenito, CC; Rogers, RD. Partitioning behavior of porphyrin dyes in aqueous biphasic systems. *Sep Sci Technol.*, 1999, 34, 1091–101. doi: 10.1080/01496399908951082.

Huddleston, JG; Willauer, HD; Boaz, KR; Rogers, RD. Separation and recovery of food coloring dyes using aqueous biphasic extraction chromatographic resins. *J Chromatogr B.*, 1998, 711, 237–44. doi: 10.1016/S0378-4347(97)00662-2.

Igarashi, L; Kieckbusch, T; Franco, T. Mass transfer in aqueous two-phases system packed column. *J Chromatogr B.*, 2004, 807, 75–80. doi: 10.1016/j.jchromb.2004.02.044.

Ivetic, DZ; Sciban, MB; Vasic, VM; Kukic, DV; Prodanovic, JM; Antov, MG. Evaluation of possibility of textile dye removal from wastewater by aqueous two-phase extraction. *Desalin Water Treat.*, 2013, 51, 1603–8. doi: 10.1080/19443994.2012.714650.

Kim, WS; Hirasawa, I; Kim, WS. Effects of experimental conditions on the mechanism of particle aggregation in protein precipitation by polyelectrolytes with a high molecular weight. *Chem Eng Sci.*, 2001, 56, 6525–6534.

Klomklao, S; Benjakul, S; Visessanguan, W; Simpson, BK; Kishimura, H. Partitioning and recovery of proteinase from tuna spleen by aqueous two-phase systems. *Process Biochem.*, 2005, 40, 3061–3067.

Li, N; Wang, Y; Xu, K; Huang, Y; Wen, Q; Ding, X. Development of green betaine-based deep eutectic solvent aqueous two-phase system for the extraction of protein. *Talanta.*, 2016, 152, 23–32. doi: 10.1016/j.talanta.2016.01.042.

Li, S; Cao, X. Enzymatic synthesis of Cephalexin in recyclable aqueous two-phase systems composed by two pH responsive polymers. *Biochem Eng J.*, 2014, 90, 301–6. doi: 10.1016/j.bej.2014.06.002.

Madhusudhan, M; Raghavarao, K; Nene, S. Integrated process for extraction and purification of alcohol dehydrogenase from Baker's yeast involving precipitation and aqueous two phase extraction. *Biochem Eng J.*, 2008, 38, 414–20. doi: 10.1016/j.bej.2007.08.007.

Mageste, AB; de Lemos, LR; Ferreira, GMD; da Silva, MCH; da Silva, LHM; Bonomo, RCF; Minim, LA. Aqueous two-phase systems: an efficient, environmentally safe and economically viable method for purification of natural dye carmine. *J Chromatogr A.*, 2009, 1216, 7623–9. doi: 10.1016/j.chroma.2009.09.048.

Mamo, G; Hatti-Kaul, R; Mattiasson, B. A thermostable alkaline active endo-β-1-4-xylanase from Bacillus halodurans S7: purification and characterization. *Enzyme Microb Technol.*, 2006, 39, 1492–8. doi: 10.1016/j.enzmictec.2006.03.040.

Moo-Young, M. *Comprehensive Biotechnology*. Boston: Newnes, 2011.

Muendges, J; Stark, I; Mohammad, S; Górak, A; Zeiner, T. Single stage aqueous two-phase extraction for monoclonal antibody purification from cell supernatant. *Fluid Phase Equilib.*, 2015, 385, 227–36. doi: 10.1016/j.fluid.2014.10.034.

Nestola, P; Peixoto, C; Silva, RR; Alves, PM; Mota, JP; Carrondo, MJ. Improved virus purification processes for vaccines and gene therapy. *Biotechnol Bioeng.*, 2015, 112, 843–57. doi: 10.1002/bit.25545.

Oliveira, LA; Sarubbo, LA; Porto, ALF; Takaki, GM; Tambourgi, EB. Partition of trypsin in aqueous two-phase systems of poly (ethylene glycol) and cashew-nut tree gum. *Process Biochem.*, 2002, 38, 693–699.

Palomares, RM. Practical application of aqueous two-phase partition to process development for the recovery of biological products. *Journal of Chromatography B.*, 2004, 807, 3-11.

Pesso, A; Vahan, KA. *Purificación de Produtos Biotecnológicos. Cap.*, 2005, 8, Ed. Manole Ltda. ISBN: 852042032x. pp. 34- 260. [*Purification of Biotechnological Products. Cap*]

Raghavarao, K; Ranganathan, T; Srinivas, N; Barhate, R. Aqueous two phase extraction—an environmentally benign technique. *Clean Technol Envir.*, 2003, 5, 136–41. doi: 10.1007/s10098-003-0193-z.

Rahimpour, F; Hatti-Kaul, R; Mamo, G. Response surface methodology and artificial neural network modelling of an aqueous two-phase system for purification of a recombinant alkaline active xylanase. *Process Biochem.*, 2016, 51, 452–62. doi: 10.1016/j.procbio.2015.12.018.

Raja, S; Murty, VR; Thivaharan, V; Rajasekar, V; Ramesh, V. Aqueous two phase systems for the recovery of biomolecules-a review. *Sci Technol.*, 2011, 1, 7–16. doi: 10.5923/j.scit.20110101.02.

Rajagopalu, D; Show, PL; Tan, YS; Muniandy, S; Sabaratnam, V; Ling, TC. Recovery of laccase from processed *Hericium erinaceus* (Bull.: Fr) Pers. fruiting bodies in aqueous two-phase system. *J Biosci Bioeng.*, 2016. doi:10.1016/j.jbiosc.2016.01.016.

Ramakrishnan, V; Goveas, LC; Suralikerimath, N; Jampani, C; Halami, PM; Narayan, B. Extraction and purification of lipase from *Enterococcus faecium* MTCC5695 by PEG/phosphate aqueous-two phase system (ATPS) and its biochemical characterization. *Biocatal Agric Biotechnol.*, 2016, 6, 19–27.

Rogers, RD; Griffin, ST. Partitioning of mercury in aqueous biphasic systems and on ABEC resins. *J Chromatogr B.*, 1998, 711, 277–83. doi: 10.1016/S0378-4347(98)00087-5.

Saxena, R; Sheoran, A; Giri, B; Davidson, WS. Purification strategies for microbial lipases. *J Microbiol Methods.*, 2003, 52, 1–18. doi: 10.1016/S0167-7012(02)00161-6.

Seyrek, E; Dubin, PL; Tribet, C; Gamble, EA. Ionic strength dependence of protein polyelectrolyte interactions. *Biomacromolecules.*, 2003, 4, 273–82.

Show, PL; Ling, TC; Lan, CWJ; Tey, BT; Ramanan, RN; Yong, ST; Ooi, CW. Review of microbial lipase purification using aqueous two-phase systems. *Curr Org Chem.*, 2015, 19, 19–29. doi: 10.2174/1385272819666141107225446.

Soohoo, JR; Walker, GM. Microfluidic aqueous two phase system for leukocyte concentration from whole blood. *Biomed Microdevices.*, 2009, 11, 323–9. doi: 10.1007/s10544-008-9238-8.

Su, CK; Chiang, BH. Partitioning and purification of lysozyme from chicken egg white using aqueous two-phase system. *Process Biochem.*, 2006, 41, 257–63.

Toner, M; Irimia, D. Blood-on-a-chip. *Annu Rev Biomed Eng.*, 2005, 7, 77. doi: 10.1146/annurev.bioeng.7.011205.135108.

Walter, H; Johansson, G. *Aqueous Two-Phase Systems*. Berlin: Elsevier, 1994.

Walter, H; Johanson, G. Partitioning in aqueous two-phase systems: an overview. *Analytical Biochemistry.*, 1986, 155, 215-242.

Willauer, HD; Huddleston, JG; Griffin, ST; Rogers, RD. Partitioning of aromatic molecules in aqueous biphasic systems. *Sep Sci Technol.*, 1999, 34, 1069–90. doi: 10.1080/01496399908951081.

Xu, Y; He, G; Li, JJ. Effective extraction of elastase from Bacillus sp. fermentation broth using aqueous two-phase system. *J Zhejiang Univ Sci B.*, 2005, 6, 1087. doi: 10.1631/jzus.2005.B1087.

Yang, H; Gurgel, PV; Carbonell, RG. Purification of human immunoglobulin G via Fc-specific small peptide ligand affinity chromatography. *J Chromatogr A.*, 2009, 1216, 910–8. doi: 10.1016/j.chroma.2008.12.004.

Zaslavsky, BY. *Aqueous two-phase partitioning: physical chemistry and bioanalytical applications.* Boca Raton: CRC Press, 1994.

Zaslavsky, BY. *Aqueous two-phase partitionig physical chemistry and bioanalytical applications.* Edited by Marcel Dekker, Inc. New York., 1995, 212-223.

In: Aqueous Two-Phase Systems   ISBN: 978-1-53614-241-9
Editor: V. Anatolijs Xanthopoulos   © 2018 Nova Science Publishers, Inc.

Chapter 2

AQUEOUS TWO-PHASE SYSTEM APPLIED IN THE PURIFICATION OF LIPASES

Thamarys Scapini[1], Aline Frumi Camargo[1],
Simone Kubeneck[1], Caroline Dalastra[1],
Charline Bonatto[1], Gislaine Fongaro[1]
and Helen Treichel[1,]*
Laboratory of Microbiology and Bioprocess,
Federal University of Fronteira Sul, Erechim, RS, Brazil

ABSTRACT

Lipases are important biocatalysts belonging to the group of hydrolases, presenting the ability to catalyze total or partial hydrolysis of triacylglycerol to diacylglycerol, monoacylglycerol, glycerol and even free fatty acids. In general, lipases are applied in a variety of processes due to their affinity with a wide range of substrates, as well as their thermal, pH and organic solvents stability. In the production of enzymes using biotechnological processes, such as lipases, the crude enzymatic extract obtained is composed of an aqueous mixture of cells, extra and intracellular products, substrates and unconverted components. After the

---

* Corresponding Author Email: helentreichel@gmail.com.

fermentation or extraction, depending on its application, enzyme purification operations are required, which should be included in the final cost of the product. Processes that allow the recovery and purification of enzymes efficiently and at low cost are extremely desirable and represent a research area of great interest. Aqueous biphasic systems are based on a liquid-liquid fractionation technique and are widely studied for the recovery and purification of enzymes, such as lipases. This technique involves the construction of extraction systems that can be formed by: (i) two polymers; (ii) a polymer and a salt; (iii), an ionic liquid and a salt or (iv) an alcohol having a low molar mass and a salt mixed over a concentration limit, thereby forming two immiscible phases. This method is widely used in protein purification because it minimizes the denaturation and/or loss of biological activity due to the high water content and low interfacial tension of the systems, which protect the proteins. In addition, two-phase aqueous systems are applicable in real scale and allow recycling of the reagents used in the process. Its high water content implies in high biocompatibility and low interfacial tension, minimizing the degradation of biomolecules. In this way, the present chapter will present the main forms of purification of lipases from aqueous biphasic systems, addressing and discussing the fundamentals for the formation, use and optimization of these systems applied in the biotechnological context. For this the following items will be addressed: i) Introduction; ii) Basis of aqueous-biphasic systems; iii) Aqueous-two-phase systems for enzymatic purification; iv) Purification of lipases; v) Prospects and combined uses of aqueous two-phase systems.

Keywords: enzymatic purification, biotechnology, combined techniques

INTRODUCTION

Aqueous Two-Phase System (ATPS) consists of the mixture of components partially miscible, two-phase formers in solvent, usually water, being based on polymer-polymer systems (usually polyethylene glycol (PEG) and dextran) and polymer-salt systems (e.g., phosphate, sulfate or citrate) (Iqbal et al., 2016; Cen et al., 2018; Lv; Zheng, 2018). Other systems types include ionic liquids (Yang et al., 2018) and short-chain alcohols (Cen et al., 2018).

The most common biphasic systems polymer/polymer and polymer/salt have been study for more than five decades by several researchers (Iqba et al., 2016), being in 1896 discovered by Martinus Beijerink, that while mixing gelatin and agar observed the formation of two different phases, which could be considered an ATPS (Albertsson, 1970), because when two polymers are mixed in a common solvent, the tendency is the formation of two distinct phases that occur due to the low entropy of the system and the low molar concentration of the polymers in solution (Baskir; Hatton and Suter, 1989).

This system is a new process method, considered a green technology, having the advantage of being non-toxic, non-flammable, easy to scale-up and low cost (Yongqiang et al., 2016), therefore have been used in a variety of applications, such as extraction of heavy metal (Hamta and Dehghani, 2017), treatment of textile industry effluents (Borges et al., 2016), bioproducts purifications (Panas et al., 2017), extraction and separation of compounds (Zhang et al., 2016), extraction of food compounds (Shao et al., 2014) and enzyme purification (Souza et al., 2015).

Currently, ATPS has been widely used as an alternative to conventional methods of protein purification, being widely used in the area of bioprocesses for the recovery and purification of biological products (Cen et al., 2018), including enzyme purification as an alternative to the extraction process with organic solvents. Because of the high water content in both immiscible phases, this system favor enzymatic purification by providing an environment with soft conditions for proteins or for living cells in the course of the process, being widely used in the processes of extractive fermentation to separate products and limit enzymatic inhibition (Vobecká et al., 2018).

Widely used for enzymatic processes, ATPS is cited in current literature in processes involving several enzymes such as pectinases (Wolf-Márquez et al., 2017), collagenase (Wanderley et al., 2017), laccases (Blatkiewicz et al., 2018), peroxidases (Targovnik; Cascone and Miranda, 2012; Lucena et al., 2017) and lipases (Duarte et al., 2015; Souza et al., 2015; Carvalho et al., 2017).

Being widely used in industry, lipases are an important group of enzymes that due to catalytic activity in aqueous and non-aqueous media, having several applications, such as pharmaceuticals, dairy products and detergent (Gupta; Sahai and Gupta, 2007), being significant due to their stability in organic solvents, functional capacity at extreme temperatures and pH (Geoffry and Achur, 2018). However, microbial lipase production by fermentation process typically requires a purification stage of intensive effort due to the complexity of broth and need to maintain enzyme activity (Show et al., 2015). As the traditional processes of purification (chromatographic methods, membrane separations and liquid-liquid extractions) are complex and expensive, alternative methods of purification are being exploited, for example ATPS (Lee et al., 2017a). Many studies can be found using this technique for recovery and purification of lipases using polyethylene glycol (PEG) and potassium phosphate (Carvalho et al., 2017), self-buffering ionic liquids, salts and copolymers (Lee et al., 2017a), salts (chloride) and tetrahydrofuran (Souza et al., 2015), alcohol (ethanol, 2-propanol and 1-propanol) and salt (ammonium sulfate, potassium phosphate and sodium citrate) (Ooi et al., 2009), as components for the formation of aqueous two-phase systems and have been shown to be an effective alternative for recovery of biomolecules.

Based on this, this chapter presents the main concepts involving processes of ATPS, the application of this system in enzyme purification, emphasizing lipases purification.

BASICS OF AQUEOUS TWO-PHASE SYSTEM

ATPS is a process consisting of two immiscible phases formed by the mixing of phases forming components above a critical concentration (Cen et al., 2018). The most studied systems of ATPS are based on polymer-polymer systems (usually polyethylene glycol (PEG) and dextran) and polymer-salt systems (e.g., phosphate, sulphate or citrate). Polymer-polymer systems presents advantages over the other supracited systems because the polymer-salt systems can present economic and environmental problems associated with the large consumption of phase-forming chemicals (Saravanan et al., 2008; Iqbal et al., 2016; Cen et al., 2018; Lv and Zheng, 2018).

Aqueous two-phase system generally uses PEG and as salt, phosphate. The first is proposed due to its terminal hydroxyl groups, whereas the phosphates are the most used when the system remains at pH close to neutrality or above this (Benavides and Rito-Palomares, 2008; Mayolo-Deloisa; Trejo-Hernadez and Rito-Palomares, 2009; Blatkiewicz et al., 2018).

The interaction of polymer-polymer, polymer-salt or other, results in two phases, which are separated by the formation of an affinity binding complex by biomolecules, which induces the substances or proteins to remain in one phase and the contaminants in other phase (Figure 1) (Iqbal et al., 2016).

ATPS is an advantageous system for many reasons, notably by easily scaling up without significant changes in process efficiency; in addition, complete mixing of the two phases occurs in only a few seconds in most systems, since interfacing is fast. Another advantage of the ATPS process is the high compatibility of several proteins with the phases, making this an attractive procedure for separation and purification of proteins (Silva and Franco, 1999). In addition, to application in biomolecules purification, aqueous two-phase system can be used for extraction and separation of metals from aqueous phase and to purify industrial samples, recovering critical metals from waste (Chen et al., 2018).

The partitioning of compounds in these systems, while suggesting simplicity, is very complex and involves many factors. The interaction of a compound with each of the phases includes hydrogen bonds, hydrophobic and charge interactions and steric effects, and even more complex when it is understood that these factors are not completely independent of one another (Marcos et al., 2002).

The phase separation in the polymer-polymer system occurs due to the steric exclusion process, which causes large aggregates to form and, consequently, two polymers tend to separate into two phases. When salts are added to the medium, a similar steric exclusion process is observed between the polymer and the high salt concentration, since the salt present in the medium will capture a significant amount of water (Asenjo and Andrews, 2011). The phase separation is influenced by factors such as molecular weight, concentration of the polymers, and when in the presence of salts, the composition and concentration of the added salt (Mazzola et al., 2008). Generally, when in polymer/salt systems, the top phase contains the highest concentration of the polymer, and the bottom phase is rich in inorganic salt or dextran (Vobecká et al., 2018).

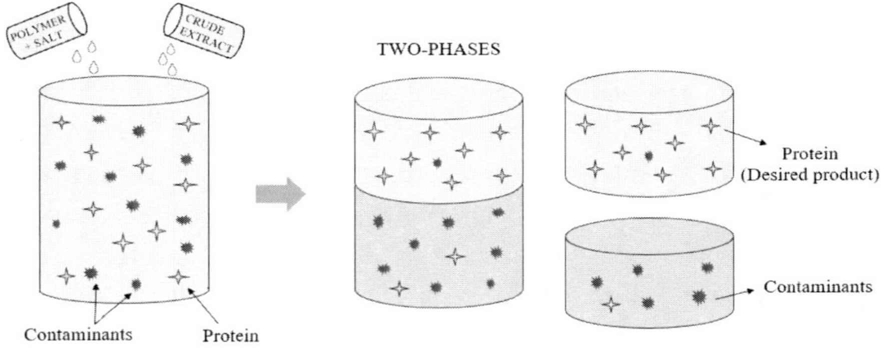

Source: Authors.

Figure 1. Scheme representative of two phases and separating the desired product.

The diagram phase (Figure 2) is like the fingerprint of an ATPS, extremely important for biomolecule partition study, because it represents the boundary that separates the region of a single phase of the region where two phases occurs (Albertsson and Tjerneld, 1994; Mazzola et al., 2008; Iqbal et al., 2016).

The main characteristic of binodal curve is the separation of two regions of distinct compositions. All compositions above the curve represent the formation of two phases in the ATPS, lower than this curve; the monophasic region is presented (Albertsson and Tjerneld, 1994).

The tie line is a bond line connecting the two-phase composition that is in the binodal curve, and all systems interconnected by these lines have equal composition of top and bottom phase equilibrium (Iqbal et al., 2016). As the polymer concentrations are reduced line is shorter, reaching the critical point, represented by C, where the compositions of the two-phase system are equal, i.e., the difference between them disappear at the critical point (Albertsson and Tjerneld, 1994). At the critical point (C), the length of the tie line is zero (Iqbal et al., 2016).

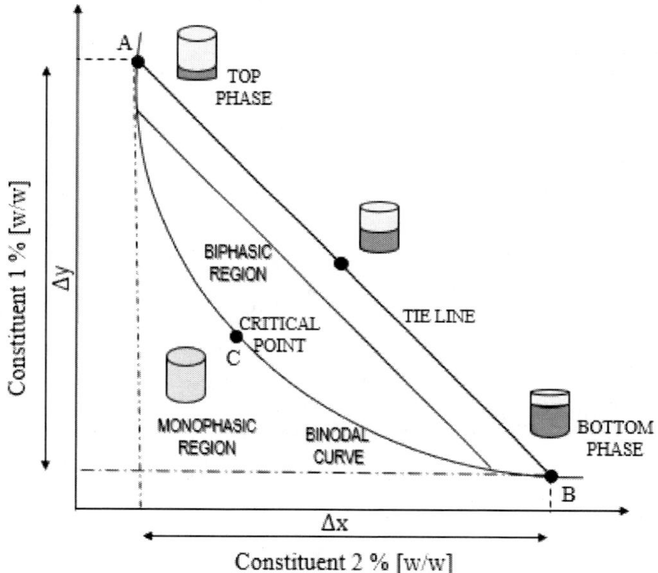

Source: Authors.

Figure 2. Representation of phase diagram. AB: tie lines; ACB: binodal curve. Constituent 1: (PEG); Constituent polymer 2: salt.

The tie line (TL) length can be estimated by using the weight ratio (Raja et al., 2011; Iqbal et al., 2016):

$$\frac{V_t \rho_t}{V_b \rho_b} = \frac{SB}{ST} \tag{1}$$

where V_t and V_b are volumes of top and bottom; ρ_t and ρ_b densities of top and bottom; SB and ST ares segments lines.

Or, the tie lines leght (TLL) can be determined using the equation below (Raja et al., 2011; Iqbal et al. 2016).

$$TLL = \sqrt{\Delta x^2 + \Delta y^2}$$

$$(2)$$

The tie lines have slope (STL), which can be determined from equation 3, where the variation in y is considered by the variation in x (Raja and Murthy, 2013).

$$STL = \frac{\Delta x}{\Delta y}$$

$$(3)$$

When the crude extract is added to the ATPS medium, the target substance and the contaminants are divided between the phases according to the affinity. By selecting the appropriate process conditions, contaminants will be confined to one phase as the target molecule divides into the other phase. To know the behavior of partitioning in the top or bottom phase, the partition coefficient (K) is determined (Baskir; Hattom and Sutter, 1989; Mazzola et al., 2008; Iqbal et al, 2016).

$$K = \frac{C_{top}}{C_{bot}}$$

$$(4)$$

where, C_{top} and C_{bot} are the concentrations of the protein in phases of top and bottom, respectively.

The partition coefficient is used to evaluate the separation of the substances in ATPS, serving as an indication of the partitioning efficiency of the substances (Mazzola et al, 2008).

The construction of the phase diagrams is performed by adding one of the system components (polymer or saline) in small aliquots and known concentration until phase separation occurs (Bamberger et al., 1983). Although a simple and rapid method, it is based on visual observation of the time of transition phase (Mazzola et al., 2008). Binodal curve can be constructed using the turbidimetric titration method, where systems with compositions in the two-phase region of the phase diagram are weighed in test tubes and water is gradually added until the turbidity of the solution disappears and the final weight is determined (Albertsson and Tjerneld, 1994).

AQUEOUS TWO-PHASES SYSTEMS FOR ENZYMATIC PURIFICATION

Enzymes are important in many industrial applications because they are efficient catalysts acting on all the chemical reactions that occur in living organisms, having the ability to make specific reactions with high rate that the chemical or physical catalysis cannot perform (Rao et al., 2014). They can be used in bioremediation by removing toxic materials and recovering valuable resources in systems, such in sewage sludge (Whiteley and Lee, 2006). In addition, to removing toxic materials and restoring the environment, enzymes can still act as indicators of the quality and state of the environment and detect pollutants before and after the restoration process (Rao et al, 2014). Table 1 presents recent research using two-phase aqueous system for enzyme purification.

Production of enzymes is usually done through fermentative processes, as it is a simple and efficient method (Blatkiewicz et al., 2018). However, for industrial application it is necessary that enzymes present a high purity and that it is possible to obtain high amounts of the purified extract, requiring the application of purification methods, generally expensive and using intensive energy (Silverio et al., 2013, Blatkiewicz et al., 2018). In this sense, the biotechnology industry has constantly sought to develop methods with high efficiency and economically viable in enzyme purification, since the process of purification of an industrial enzyme is the most important stage, accounting for at least 60% of the total cost of the process (Silverio et al., 2013; Loureiro et al., 2017).

Enzyme purification is carry out aiming at removing contaminants which may be present in the enzyme extract and which may affect in stability and efficiency enzymatic (Vijayaraghavan; Raj and Vincent, 2016). The purification method varies according to the source and the property of the enzyme to be purified and several methods have been reported in literature, including ultrafiltration (Kitaoka and Robyt, 1998; Sánchez-González et al., 1999; Guzman, Hurtado and Ospina, 2018) and chromatographic methods (Purwanto, 2016; Alici and Arabaci, 2018) and aqueous two-phase systems (ATPS), being this the most popular and viable method of enzymatic purification (Blatkiewicz et al., 2018).

**Table 1. Recent research using two-phase aqueous
system for enzyme purification**

Enzyme	ATPS compounds	PF*	Application	Reference
Collagenases	PEG + Phosphate	27.61	Industry, medicine and biotechnology	Wanderley et al. (2017)
NAD(P)H oxidase	PEG-KH2PO4 + NaCl	8.33	Regeneration of cofactors in processes catalyzed by oxidoreductases	Tongul; Kavakcioglu and Tarhan (2016)
Protease	PEG + citrate	3.95	Replacement of toxic chemicals	Rao et al. (1998) Silva et al. (2017)
Lipase	Flotation system in the fermentation process combination of aqueous two phase system	11.48	Scale-up for applied in industries	Sankaran et al. (2018a)
Peroxidase	PEG 400, 1000, 1500, 4000, 6000, 8000 + ammonium sulfate salt and ionic liquid in ATPS	19.2	Separation potential, the purification of radish peroxidase from natural source, the low cost	Lucena et al., (2017)

*Purification Factor (PF).

The use of ATPS is a very interesting alternative, being widely applied in the chemical industry for its simplicity, economical feasibility and easy of expansion without considerable loss of process efficiency (Silvério et al., 2013). In biomolecule purification techniques, ATPS offers many advantages over conventional methods, and can replace solid-liquid separation processes, offering many advantages over conventional methods, including low processing time, since phase separation usually occurs in a few seconds, low cost material, low energy consumption and high biocompatibility, since 80-95% of the weight of the system is water, which favors the conditions of purification of biomolecules, as enzymes, since they provide smooth conditions of process and avoid the rupture of the cells and consequent denaturation of the enzymes (Gavasane and Gaikar, 2003; Silvério et al. 2013; Wanderley et al., 2017).

ATPS was reported by Beijerink in 1896 and has been studied by several researchers using a combination of different polymers and different ligands for enzyme purification (Albertsson, 1896; Gavasane and Gaikar, 2003). The combination of different polymer systems and the addition of salts and other compounds, results in different purification factors for different industrial applications, some examples being shown in Table 1.

Several factors influence ATPS partitioning process, and these are studied and optimized according to the extract which the enzymes are inserted, since the mechanisms by which the affinity of the proteins with the ATPS can causes separation in the top or bottom phase (Iqbal et al., 2016). When two or more factors influence the partitioning of a target molecule in the ATPS, the overall coefficient is determined from the product of several contributions, expressed in the logarithmic form of Equation (5) (Andrews; Schmidt and Asenjo, 2005; Raja et al., 2011; Grillo; Aires-Barros and Azevedo, 2014).

$$K = K_{size} \cdot K_{elec} \cdot K_{hfob} \cdot K_{aff} \cdot K_{o}$$

$$(5)$$

where K_{size}, K_{elec}, K_{hfob}, K_{aff}, and K_{o} correspond to the contribution of the overall partition coefficient by size, electrostatic forces, hydrophobicity, affinity and environmental factor, respectively.

The Molecular Weight (MW) of the polymers is the most important characteristics in ATPS. Generally, the higher the MW of the polymer the lower is polymer concentration needed to form the phases and consequently the partition coefficient (K), is lower (Iqbal et al., 2016). This characteristic was observed by Blatkiewicz et al. (2018) in a study carried out with Laccase from *C. unicolor*, where in PEG of MW 1500, 3000 and 6000 it was observed a greater affinity towards the top phase of the ATPS, already in PEG 400 the partitioning activity was observed in the bottom phase.

Considering the concentration of the polymer in the system, it was observed that the higher the concentration of the polymer, greater is the difference in properties between the phases, reflecting in the phase diagram, with longer lines (Albertsson, 1994). In polymer-salt, the partition decreases towards the polymer phase when a system with a high concentration of polymer is present, whereas in polymer-polymer systems partition decreases towards the phase with higher molecular weight (Iqbal et al., 2016).

Hydrophobicity in the protein partition influences through two main effects: phase hydrophobicity effect and salting out effect (Andrews; Schmidt; Asenjo, 2005; Asenjo and Andrews, 2011; Raja et al., 2011; Iqbal, 2016). The presence of neutral salts, such as NaCl, does not alter the protein partitioning system in ATPS (<1M) (Raja et al., 2011; Iqbal et al., 2016). However, when the saline concentration in the medium is increased (> 1M), the influence of hydrophobicity is observed, since the hydrophobic ions and of different hydrophobicides force counterions to change to the phase with greater hydrophobicity, this effect generates the so-called salting- where the biomolecules tend to migrate from the phase with the highest concentration of salt to the phase with the highest concentration of PEG (Andrews; Schmidt; Asenjo, 2005; Asenjo, Andrews, 2011; Raja et al., 2011; Iqbal, 2016).

Opposite charges attraction and pH has a great influence on the electrochemical interactions, so it is considered that the manipulation of system pH may promote the selective fractionation of solute and particles with opposite charges towards to some of the ATPS phases (Asenjo and Andrews, 2011; Benavides and Rito-Palomares, 2011). The PEG polymer, most commonly used in ATPS, has a terminal hydroxyl groups, which gives it a positive dipole moment, thus, the use of pH above the protein isoeletric point will induce a greater affinity with the phase of higher concentration of PEG and consequently the partition coefficient will be higher (Benavides and Rito-Palomares, 2011; Raja et al., 2011). Generally, at high pH (>6) the protein is negatively charged and therefore the partition coefficient increases with pH increasing (Saravanan et al., 2008).

Temperature is a factor that influences the composition of two phases in an ATPS and the partition through viscosity, surface tension and density (Iqbal et al., 2016). Generally, two characteristics are observed with changes in temperature: in polymer-polymer systems the phase separation at low concentrations of the polymer is achieved when maintained at low temperatures. The opposite is observed in polymer-salt systems, where higher temperatures reduce the solubility of PEG in water, and consequently the phase separation can be obtained in lower concentrations (Albertsson and Tjerneld, 1994; Iqbal et al., 2016).

Despite the understanding of the factors that influence ATPS, selectivity in protein partitioning is a feature that needs further investigation, since it will lead to the ability to predict the partition of target proteins, many times produced and found with other proteins and heterogeneous and complex extracts, sometimes being partitioned in the upper phase, and at other times in the lower phase, resulting, in some cases, in low reproducibility (Silva and Franco, 1999).

PURIFICATION OF LIPASES

Lipases (EC. 3.1.1.3) belong to a class of hydrolases that catalyze the hydrolysis of insoluble triacylglycerols to generate free fatty acids, diacylglycerols, monoacylglycerols and glycerol in an oil-water interface (Geoffry and Achur, 2018a; Treichel et al., 2010). They can also catalyze reactions of esterification, transesterification and interesterification in organic solvents (Villeneuve et al., 2000).

These enzymes can be obtained from several species of bacteria, fungi, yeast, plants and animals. Lipases from microorganisms are important for biotechnological applications due to its functional capacity in extreme temperatures and a wide range of pH, stability in organic solvents, chemoselectivity, rhegium-selectivity and enantio-selectivity (Geoffry and Achur, 2018a). In addition, many agro-industrial wastes containing oil as carbon source, can be used as substrate for microorganisms producers of lipases (Table 2), which is advantageous economically, because it reduces the costs of enzyme production (Thakur, 2012).

For this reason, microbial lipases have attracted the attention of various industries for applications in organic chemical processing, formulations of detergents, synthesis of biosurfactants, paper manufacturing, cosmetics and pharmaceutical processing (Gupta, Sahai and Gupta, 2007) and have also been studied for application in generation of biofuels and treatment of effluents and contaminated soils with high levels of oils and derivatives (Table 2) (Hama, Noda and Kondo, 2018; Margesin, Zimmerbauer and Schinner, 1999; Okoro et al., 2018).

For the production of microbial lipases is necessary to perform a stage of fermentation. This process can be carried out using submerged or solid state fermentation. The submerged fermentation is a process of cultivation of microorganisms in an environment that is in liquid form. Among the advantages of this system, the greater homogeneity of the culture medium and easy control variables, such as temperature and pH (Geoffry and Achur, 2018a), can be highlighted. However, the need to use oil in liquid form makes the process expensive, since the oil is a component of high cost. (Sethi, Nanda and Sahoo, 2016). In solid state fermentation (SSF) microorganisms are grown in solid substrates without the presence of free water (Geoffry and Achur, 2018a). The SSF is preferred due to its simplicity, low level of catabolic repression and inhibition of the final product, low wastage of water, better recovery of the product and high quality production (Lonsane et al., 1985). In addition, purified lipases produced by SSF system have significantly greater thermal stability to that produced by SMF (Sethi, Nanda and Sahoo, 2016). Another advantage associated with the production of lipases through SSF is the possibility of using waste as a substrate for cultivation of microorganisms' producers of lipases (Treichel et al., 2010). Filamentous fungi are pointed out as the most suitable for bioconversion of solid substrates due to its growth mode, good adaptability in low water content and conditions of high osmotic pressure (Sethi, Nanda and Sahoo, 2016).

After the fermentation, lipase is a complex extract containing a series of other compounds that are not of interest. For this reason, the production of lipases through microbial fermentation normally requires one or more steps of purification aiming at obtaining the enzyme with high purity, which is essential for their industrial application (Show et al., 2015).

For this reason, initially is usually performed a pre-purification, which consists in the removal of cells from the fermentation broth, either by centrifugation or filtration. The culture broth free of cells is concentrated by precipitation, ultrafiltration or extraction with organic solvents (Saxena et al., 2003). Purification steps used to start by a precipitation process and go to until the column chromatography, depending on the nature of the protein and the desired level of purification (Javed et al., 2018). According to Saxena et al. (2003) 80% of purification systems presented in the literature uses precipitation followed by chromatographic methods, such as gel filtration and affinity chromatography. Approximately 60% of the studies evaluated the precipitation with ammonium sulphate and only 35% used ethanol, acetone or a hydrochloric acid.

Separations of high resolutions, such as chromatography, have high costs of installation, maintenance and operation due to the complexity of the system, which makes the use at industrial level economically unviable. So, industries seek economically feasible technologies, fast and capable of scale up. As an alternative, new technologies have been developed and are widely reported in the literature, such as aqueous two-phase systems (Saxena et al., 2003).

Among the studies that report lipases purification using ATPS. Lee et al. (2017b) used polypropylene glycol presenting average molecular weight of 400 g/mol (PPG 400), and ionic liquid, composed of naturally derived cholinium cation and anion obtained from biological buffer, i.e., BES, designated the [Ch][BES], for purification of lipase from *Burkholderia cepacia* ST8. This study obtained a purification factor of 17.96±0.32 and a recovery yield of 99.30% ± 0.03.

Derived lipase from *Yarrowia start* IMUFRJ 50682 was purified by Carvalho et al. (2017) through an aqueous two-phase system (ATPS) composed of polyethylene glycol (PEG 1500) and potassium phosphate, achieving a purification factor exceeding 40 and increase the thermostability of the enzyme (up to 50ºC). Other studies used PEG and potassium phosphate as components for the ATPS for purification of lipase: Zhou et al. (2013) purified porcine pancreatic lipase, Barbosa et al. (2011) performed a purification of lipase produced by *Bacillus* sp. ITP-001 in submerged fermentation and Ramakrishnan et al. (2016) used this system for purification of lipase from *Enterococcus faecium* MTCC5695.

Aqueous biphasic system formed by 20% PEG 2000 and 12% ammonium sulphate, with the addition of 5% of sodium carbonate at pH 8 to 30% of gross load, was used by Anvari (2015) for the purification of the lipase from *Rhizopus microsporus*. The author got a recovery yield of 92.3% and a factor of purification in the upper phase of 19.8 when using this system.

Souza et al. (2015) analyzed the formation of ATPS based on cholinium-based salts (cholinium chloride, cholinium bitartrate and cholinium dihydrogencitrate) and tetrahydrofuran (THF) for the purification of lipase from *Bacillus* sp. ITP-001, produced by submerged fermentation and found that the optimal conditions (40 wt% of THF and 30 wt% of cholinium bitartrate at 25°C) led to a purification factor of 130.1 ± 11.7 times and lipase yield of 90.0 ± 0.7%.

Another system with potential application due to easy recovery of the enzyme, but little explored is the purification through aqueous two-phase system formed by alcohol/salt. Ooi et al. (2009) used this system to retrieve the lipase derived from *Burkholderia pseudomallei*. The authors assessed nine different systems composed of an upper phase on the basis of alcohol (ethanol, 2-propanol and 1-propanol) and a lower phase on the basis of salt (ammonium sulphate, potassium phosphate and sodium citrate). They obtained a better efficiency of purification in an ATPS composed by 16% (p/p) of 2-propanol and 16% (p/p) of phosphate in the presence of 4.5% (p/v) of sodium chloride. The yield of the system was 99% and the purification factor of 13.5. More recently, ATPS system composed of alcohol (1-propanol)/salt (ammonium sulphate) was investigated by Sankaran et al., (2018b). An advantage of the use of the ATPS composed of alcohol/salt lies in the possibility of recycling the alcohol through evaporation and crystallisation of salt (Sankaran et al., 2018b).

Table 2. Microorganisms and substrates cited in the recent literature, as potential lipase producers for different applications

Microorganism	Source	Substrate	Application	Reference
Start yarrowia	Yeast	Synthetic medium containing olive oil	biosynthesis of l-ascorbyl palmitate	Ping et al. (2018)
Escherichia coli BL21 (3)	Bacterial	Pseudostem and frond of banana plant (Musa sp.)	Lipase Production	Chai et al. (2018)
Aspergillus niger	Fungal	Canola cake	Dairy wastewater treatment	Golunski et al. (2017)
Staphylococcus aureus	Bacterial	Synthetic medium containing olive oil	Detergent formulations	Bacha et al. (2018)
Fusarium solani NFCCL strain 4084	Fungal	Palm oil mill effluent	Lipase Production	Geoffry and Achur (2018b)
Aspergillus niger	Fungal	Canola meal	Pretreatment technique of oil	Mulinari et al. (2017)
Baccilus licheniformis KM12	Bacterial	Synthetic medium containing olive oil	Biodiesel production	Malekabadi, Badoei-Dalfard and Karami (2018)

Other option shown in the literature is lipase purification through ATPS using polymer/polymer. The most commonly polymers used are polyethylene glycol (PEG) and dextran (Saxena et al., 2003). Ooi et al. (2011) evaluated a purification of lipase produced by *Burkholderia pseudomallei* by ATPS composed by 9.6% (w/w) PEG 8000 and 1.0% (w/w) dextran T500 and achieves a yield of 92.1%. However, due to the high cost of the dextran, the ATPS composed of polymer/polymer is generally perceived as less attractive in comparison with the polymer/salt (Show et al., 2015).

Finally, systems composed by polymers and salt, PEG/phosphate and potassium PEG/magnesium sulphate are the most commonly used for purification of microbial lipases (Saxena et al., 2003).

PROSPECT AND COMBINED USES OF AQUEOUS TWO-PHASES SYSTEM

In pioneering studies, ATPS was based on polymer/polymer or polymer/salt combinations. More recently, studies using ionic liquids associated to ATPS appears as a method widely used mainly to obtain biomolecules that need to be removed from solutions containing more than one molecule in suspension, obtaining and/or purification of enzymes and proteins are examples of the application of this method (Li et al., 2012; Ventura et al., 2012; Zhang et al., 2017).

Ionic liquids (ILs) are non-flammable solvents composed entirely of ions thatpresent negligible volatility and comprehensive solubility for organic and inorganic compounds (Wilkes et al., 2004; Lucena et al., 2017; He et al., 2018). The combination of ATPS and ILs creates an efficient and lowcost purification and separation system (An et al., 2017), gathering the advantages of both (He et al., 2018). The IL-based ATPS have already been used for the extraction and/or separation of several compounds (Marques et al., 2013) as amino acids (Ventura et al., 2009), proteins (Du et al., 2007), enzymes (Tzeng et al., 2008), among others.

Despite its advantages, the process of IL synthesis is complex and expensive, which limits its application and development on an industrial scale, so the search for alternative and simpler solvents is essential (Zeng et al., 2014). According to Zhang et al. (2017) due to the search for alternative methods and that present as environmentally viable arise the Deep Eutectic Solvent (DES), in its main connections consist of hydrogen bonds (Zeng et al., 2014).

In 2003, studies performed by Abbott et al. demonstrated, for the first time, that the mixture of choline with urea can produce eutectic mixtures, which at room temperature would be liquid. This type of mixture was defined as a Deep Eutectic Solvent (DES), but it is also called IL analogs, because they have similar properties (Zeng et al., 2014; Baghlani and Sadegi, 2018). The great advantage of DES is that it is ecologically correct, biodegradable, cheap and easy to apply (Abbott et al., 2003; Zhang et al., 2017; Baghlani and Sadegi, 2018).

Alternatives are being sought with a focus on safe methodologies aimed at reducing both human health and environmental risks. Based on this assumption, researchers analyze the substitution of toxic reagents for those viable ones, with low cost and that are considered friendly of the environment, for that, new approaches are studied for purification and/or extraction of biomolecules of interest (Welton, 1999; Pena-Pereira; Lavilla and Bendicho, 2010; Armenta; Garrigues and De la Guardia, 2015; Li and Row, 2016). Some of these approaches include: deep eutectic solvent (DES), DES-salts, DES-polymers.

One of the determining factors in biotechnological processes is the viability of large-scale production. In this sense, the methodology used must prove attractive in the industrial environment. For this reason, properties such as low cost, ease of operation and reusability make the process more sustainable. Biotechnological processes related to the separation and/or purification of biomolecules, for example purification of enzymes, are frequently carried out in food industries (Gaikaiwari et al., 2012) and environmental research, among these purification methods, most of them have a very low operating cost, besides the need for skilled workforce for analysis, is the case of the chromatographic technique (Sankaran et al., 2018a). Many of these techniques require multiple unit operations, long extraction times, and as consequence inactive the biomolecule (Sousa et al., 2018).

The search for cleaner methods of enzymatic purification aiming at environmentally friendly and present efficient alternatives techniques are a constant search by researchers as well as in industrial environments, especially looking for the increment of the final product (Sankaran et al.,

2018a), therefore Table 3 brings a brief compilation of trends in this area, focusing on methods combined with the purpose of enhancing their applications.

Table 3. Trends of methods combined with the aqueous two-phases system, presenting the combination, application of the study and their authors

Combination	Application	Reference
Liquid biphasic flotation process is an alternative method that involves the combination of aqueous two-phase system (ATPS) and solvent sublation (SS) system	Lipase production and separation, the novel method able to increase the productivity at a shorter period.	Sankaran et al. (2018a)
Ionic liquids and self-buffering	Extraction and purification of the enzyme α-chymotrypsin	Gupta; Taha and Lee (2016)
Aqueous two-phases system (ATPS) based on ionic liquids	Extraction of pesticides of different polarity (acetamiprid, imidaclopride, simazine, linuron and tebufenozide)	Dimitrijevic et al. (2017)
Based in seven polymers (PEG), ammonium sulfate salt and protic ionic liquids	Separation and purification of radish peroxidase from natural source	Lucena et al. (2017)
Alcohol combination with salt ATPS	Recovering enhanced green fluorescent protein from a clarified lysate of recombinant *E. coli*	Lo et al. (2018)
Employed a biphasic system consisting of dilute HCl (aqueous) and DCM (dichloromethane organic extraction solvent)	Deconstruction of agriculture residue (rice straw) to levulinic acid with a viable method by achieving maximum product recovery	Kumar et al. (2018)
Twelve kinds of salts were tested and combined with a new deep eutectic solvent (DES), which was composed of tetrabutylammonium bromide (TBAB) and PPG 400	Investigation of pigments hydrophobicity on extraction efficiency	Zhang et al. (2018)

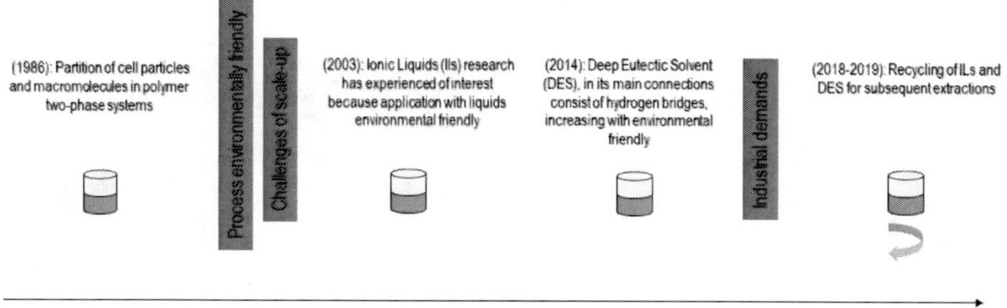

Source: Authors.

Figure 3. Flowchart showing trend in use the aqueous two-phases system technique.

Methods of separation and/or purification biomolecules through ATPS are boosten by research on economically and environmentally viable methods, and in this case have a tendency to be catalyzed by ionic liquids and deep eutectic solvent, both are potential beneficial for the environment and are strong substituents for volatile organic compounds (Shukla; Pandey and Pandey, 2017).

Based on the idea that these systems are considered low cost and environmental friendly, some current researchers present studies on the recycling of IL-based ATPS for subsequent extractions (An et al., 2017). According to Almeida et al. (2016), an aqueous salt solution, in this case aluminum sulphate ($Al_2(SO_4)_3$), may be employed in the recycling of used ILs. In their study, they proved that recycled IL can be used by four extraction cycles without loss of ATPS performance, ensuring system sustainability (Shukla; Pandey and Pandey, 2017). Pereira et al. (2015) also verified in their study that the IL-based ATPS can be recycled and reused by three cycles of extraction without loss of process efficiency. Temporal evolution of ATPS technique is demonstrated in Figure 3.

Lastly, investigations in the use of ATPS and the combination of other purification techniques have been studied, since this type of modification already presented significant yields of both purification and recovery. In this way, the affinity of the ATPS with other systems should be evaluated as an application tendency for the purification of various biological products, presenting high biotechnological potential.

REFERENCES

Abbott, A. P.; Capper, G.; Davies, D. L.; Rasheed, R. K.; Tambyrajah, V. (2003). Novel solvent properties of choline chloride/urea mixtures. *Chemical Communications*, 70-71.

Albertsson, P. (1970). Partition of cell particles and macromolecules in polymer two-phase systems. *Advances in protein chemistry*, *24*, 309-341.

Albertssonn, P. A. (1986). *Partition of Cell Particles and Macromolecules*. New York: Wiley-Interscience.

Albertsson, P. A. & Tjerneld, F. (1994). Phase diagrams. In ..., *Methods in Enzymology*. City, ST: Publisher.

Alici, E. H.; Arabaci, G. (2018). A novel serine from strawberry (*Fragaria ananassa*): Purification and biochemical characterization. *Biological Macromolecules*, *114*, 1295-1304.

Almeida, H. F. D.; Freire, M. G.; Marrucho, I. M. (2016). Improved Extraction of Fluoroquinolones with Recyclable Ionic-Liquid-based Aqueous Biphasic Systems. *Green Chemistry*, *18*, 2717-2725.

An, J.; Trujillo-Rodríguez, M. J.; Pino, V.; Anderson, J. L. (2017). Non-conventional solvents in liquid phase microextraction and aqueous biphasic systems. *Journal of Chromatography A*, *1500*, 1-23.

Andrews, B. A.; Schmidt, A. S.; Asenjo, J. A. (2005). Correlation for the Partition Behavior of Proteins in Aqueous Two-Phase Systems: Effect of Surface Hydrophobicity and Charge. *Biotechnology and Bioengineering*, *90*, 380-390.

Anvari, M. (2015). Extraction of lipase from Rhizopus microsporus fermentation culture by aqueous two-phase partitioning. *Biotechnology & Biotechnological Equipment*, *29*, 723-731.

Armenta, S.; Garrigues, S.; De la Guardia, M. (2015). The role of green extraction techniques in Green Analytical Chemistry. *Trends in Analytical Chemistry*, *71*, 2–8.

Asenjo, J. A.; Andrews, B. A. (2011). Aqueous two-phase systems for protein separation: a perspective. *Journal of Chromatography A*, *1218*, 8826-8835.

Bacha, A. B.; Al-Assaf, A.; Moubayed, N. M.S; Abid, I. (2018). Evaluation of the novel thermo-alkaline lipase Staphylococcus aureus for application in detergent formulations. *Saudi Journal of Biological Sciences, 25,* 409-417.

Baghlani, M.; Sadeghi, R. (2018). Thermodynamics investigation of phase behavior of deep eutectic solvents-polymer aqueous biphasic systems. *Polymer, 143,* 115-128.

Barbosa, J. M. P.; Souza, R. L.; Fricks, A. T. Zanin, G. M.; Soares, C. M. F; Lima, A. S. (2011). Purification of lipase produced by a new source of Bacillus in submerged fermentation using an aqueous two-phase system. *Journal of Chromatography B, 879,* 3853-3858.

Baskir, J. N.; Hatton, T. A.; Suter, U. W. (1989). Protein Partitioning in Two-Phase Aqueous Polymer Systems. *Biotechnology and Bioengineering, 34,* 541-558.

Benavides, J.; Rito-Palomares, M. (2008) Practical experiences from the development of aqueous two-phase processes for the recovery of high value biological products. *Journal of Chemical Technology Biotechnology, 83,* 133–142.

Benavides, J.; Rito-Palomares, M.; Ansejo, J. A. (2011). Aqueous Two-Phase Systems. *Comprehensive Biotechnology* (697-713).

Blatkiewicz, M.; Antecka, A.; Boruta, T.; Górak, A.; Ledakowicz, S. (2018). Patitioning of laccases derived from *Cerrena unicolor* and *Pleurotus sapidus* in polyethylene glycol - phosphate aqueous two-phase systems. *Process Biochemistry, 67,* 165-174.

Borges, G. A.; Silva, L. P.; Penido, J. A.; Lemos, L. R. de; Mageste, A. B.; Rodrigues, G. D. (2016). A method for dye extraction using an aqueous two-phase system: effect of co-occurrence of contaminants in textile industry wastewater. *Journal of Environmental Management, 183,* 196-203.

Carvalho, T.; Finotelli, P. V.; Bonomo, R. C. F.; Franco, M.; Amaral, P. F. F. (2017). Evaluating aqueous two-phase systems for Yarrowia lipolytica extracellular lipase purification. *Process Biochemistry, 53,* 259–266.

Cen, L. S.; Ramanan, R. N.; Ti, T. B.; Siang, T. W.; Loke, S. P.; Chuan, L. T.; Wei, O.C. (2018). Purification of the recombinant enhanced green fluorescent protein from Escherichia coli using alcohol + salt aqueous two-phase systems. *Separation and Purification Technology, 192*, 130-139.

Chai, S. Y.; Abbasiliasi, S.; Lee, C.K.; Ibrahim, T. A.T.; Kadkhodaei, S.; Mohamed, M.S.; Hashim, R.; Tan, J. S. (2018). Extraction of fresh banana waste juice non-cellulosic and non-food renewable feedstock is direct lipase production. *Renewable Energy, 126*, 431-436.

Chen, Y.; Wang, H.; Pei, Y.; Wang, J. (2018). A green separation strategy for neodymium (III) from cobalt (II) and nickel (II) using an ionic liquid-based aqueous two-phase system. *Talanta, 182*, 450-455.

Dimitrijevic, A.; Ignjatovic, L.; Tot, A.; Vranes, M.; Zec, N.; Gadzuric, S.; Trtic-Petrovic, T. (2017). Simultaneous extraction of pesticides of different polarity applying aqueous biphasic systems based on ionic liquids. *Journal of Molecular Liquids, 243*, 646-653.

Duarte, A. W. F.; Lopes, A. M.; Molino, J. V. D.; Pessoa, A.; Sette, L. D. (2015). Liquid-liquid extraction of lipase produced by psychrotrophic yeast *Leucosporidium scottii* L117 using aqueous two-phase systems. *Separation and Purification Technology, 156*, 215-225.

Du, Z.; Yu, Y. L.; Wang, J. H. (2007). Extraction of Proteins from Biological Fluids by Use of an Ionic Liquid/Aqueous Two-Phase System. *Chemistry: A Europe Journal, 13*, 2130-2137.

Gaikaiwari, R. P., Wagh, S. A., Kulkarni, B. D. (2012). Efficient lipase purification using reverse micellar extraction. *Bioresource Technology, 108*, 224–230.

Gavasane, M. R.; Gaikar, V. G. (2003). Aqueous two-phase affinity partitioning of *Pennicillin acylase* from *E. coli* in presence of PEG-derivatives. *Enzyme and Microbial Technology, 32*, 665-675.

Geoffry, K.; Achur, R. N. (2018a) Screening and production of lipase from fungal organisms. *Biocatalysis and Agricultural Biotechnology, 14*, 241-253.

Geoffry, K.; Achur, R. N. (2018b). Optimization of novel halophilic lipase production by *Fusarium solani* strain NFCCL 4084 using palm oil mill effluent. *Journal of Genetic Engineering and Biotechnology*. https://doi.org/10.1016/j.jgeb.2018.04.003.

Golunski, S. M.; Mulinari, J.; Camargo, A. F.; Venturin, B.; Baldissarelli, D. P.; Marques, C. T.; Vargas, G. D. L. P.; Colla, L. M.; Mossi, A.; Treichel, H. (2017). Ultrasound effects on the activity of *Aspergillus niger* lipases in their application in dairy wastewater treatment. *Environ Qual Manage, 27,* 95-101.

Grilo, A. L.; Aires-Barros, M. R.; Azevedo, A. M. (2014). Partitioning in Aqueous Two-Phase Systems: fundamentals, applications and trends. *Separation & Purification Reviews, 45,* 68-80.

Gupta, N.; Sahai, V.; Gupta, R. (2007). Alkaline lipase from a novel strain *Burkholderia multivorans*: statistical medium optimization and production in a bioreactor. *Process Biochemistry, 42,* 518-526.

Gupta, B. S.; Taha, M.; Lee, M. (2016). Extraction of an active enzyme by self-buffering ionic liquids: a green medium for enzymatic research. *Royal Society of Chemistry, 6,* 18567-18576.

Guzman, G. Y. F.; Hurtado, G. B.; Ospina, S. A. (2018). New dextransucrase purification process of the enzyme produced by *Leuconostoc mesenteroides* IBUN 91.2.98 based on binding product and dextranase hydrolysis. *Journal of Biotechnology, 265,* 8-14.

Hama, S.; Noda, H.; Kondo, A. (2018). How lipase technology contributes to evolution of biodiesel production using multiple feedstocks. *Current Opinion in Biotechnology, 50,* 57-64.

Hamta, A.; Dehghani, M. R. (2017). Application of polyethylene glycol based aqueous two-phase systems for extraction of heavy metals. *Journal of Molecular Liquids, 231,* 20-24.

He, A.; Dong, B.; Feng, X.; Yao, S. (2018) Extraction of bioactive ginseng saponins using aqueous two-phase systems of ionic liquids and salts. *Separation and Purification Technology, 196,* 270-280.

Iqbal, M.; Tao, Y.; Xie, S.; Zhu, Y.; Chen, D.; Wang, X.; Huang, L.; Peng, D.; Sattar, A.; Shabbir, M. A. B.; Hussain, H. I.; Ahmed, S.; Yuan, Z. (2016). Aqueous two-phase system (ATPS): an overview and advances in its applications. *Biological Procedures Online, 18,* 1-18.

Javed, S.; Azeem, F; Hussain, S.; Rasul, I.; Siddique, M.H.; Riaz, M.; Afzal, M.; Kouser, A.; Nadeem, H. (2018). Bacterial lipases: A review on purification and characterization. *Progress in Biophysics and Molecular Biology*, *132*, 23-34.

Kitaoka, M., Robyt, J. F. (1998). Large-scale preparation of highly purified dextransucrase from a high-producing constitutive mutant of *Leuconostoc mesenteroides* B-512FMC. *Enzyme Microbial Technology, 23*, 386–391.

Kumar, S.; Ahluwalia, V.; Kundu, P.; Sangwan, R. S.; Kansal, S. K.; Runge, T. M.; Elumalai, S. (2018). Improved levulinic acid production from agri-residue biomass in biphasic solvent system through synergistic catalytic effect of acid and products. *Bioresource Technology, 251*, 143-150.

Lee, S. Y.; Khoiroh, I; Coutinho, J. A. P.; Show, P. L.; Ventura, S. P. M. (2017a). Lipase production and purification by self-buffering ionic liquid-based aqueous biphasic systems. *Process Biochemistry, 63*, 221–228.

Lee, S. Y.; Khoiroh, I.; Ling, T. C.; Show, P. L. (2017b). Enhanced recovery of lipase from Burkholderia cepacia derived from fermentation broth using recyclable ionic liquid/polymer-based aqueous two-phase systems. *Separation and Purification Technology, 179*, 152-160.

Li, Z.; Liu, X.; Pei, Y.; Wang, J.; He, M. (2012). Design of environmentally friendly ionic liquid aqueous two-phase systems for the efficient and high activity extraction of proteins. *Green Chemistry, 14*, 2941-2950.

Li, X.; Row, K. H. (2016). Development of deep eutectic solvents applied in extraction and separation, *Journal of Separation Science, 39*, 3505–3520.

Lo, S. C.; Ramanan, R. N.; Tey, B. T.; Tan, W. S.; Show, P. L.; Ling, T. C.; Ooi, C. W. (2018). Purification of the recombinant enhanced green fluorescent protein from *Escherichia coli* using alcohol + salt aqueous two-phase systems. *Separation and Purification Technology, 192*, 130-139.

Lonsane, B. K.; Ghildyal, N. P.; Budiatman, S.; Ramakrishna, S. V. (1985). Engineering aspects of solid state fermentation. *Enzyme Microb. Technol, 7*, 258-265.

Loureiro, D. B.; Braia, M.; Romanini, D.; Tubio, G. (2017). Partitioning of xylanase from *Thermomyces lanuginosus* in PEG/NaCit aqueous two-phase systems: Structural and functional approach. *Protein Expression and Purification, 129,* 25-30.

Lucena, I. V.; Brandão, I. V.; Mattedi, S.; Souza, R. L.; Soares, C. M. F.; Fricks, A. T.; Lima, A. S. (2017). Use of protic ionic liquids as adjuvants in PEG-based ATPS for the purification of radish peroxidase. *Fluid Phase Equilibria, 452*, 1-8.

Lv, H.; Zheng, Y. (2018). A Newly developed tridimensional neural network for prediction of the phase equilibria of six aqueous two-phase systems. *Journal of industrial and engineering chemistry, 57,* 377-386.

Margesin, R.; Zimmerbauer, A.; Schinner, F. (1999). Soil lipase activity - a useful indicator of oil biodegradation. *Biotechnology Techniques, 13,* 859-863.

Malekabadi, S.; Badoei-dalfard, A.; Karami, Z. (2018). Biochemical characterization of a novel cold-active, halophilic andorganic solvent-tolerant lipase from *B. licheniformis* KM12 withpotential application for biodiesel production. *International Journal of Biological Macromolecules, 109,* 389-398.

Marques, C. F. C.; Mourão, T.; Neves, C. M. S. S.; Lima, A. S.; Boal-Palheiros, I.; Coutinho, J. A. P.; Freire, M. G. (2013). Aqueous biphasic systems composed of ionic liquids and sodium carbonate as enhanced routes for the extraction of tetracycline. *Biotechnology Progress, 29,* 645-654.

Mayolo-Deloisa, K.; Trejo-Hernandes, M. del R.; Rito-Palomares, M. (2009) Recovery of laccase from the residual compost of *Agaricus bisporus* in aqueous two-phase systems. *Process Biochemistry, 44,* 435–439.

Mulinari, J.; Venturin, B.; Sbardelotto, M.; Dall Agnol, A., Scapini, T.; Camargo, A. F.; Baldissarelli, D. P.; Modkovski, T. A.; Rosseto, V.; Dalla Rosa, C. Reichert Jr. F. W., Golunski, S. M.; Vieitez, I.; Vargas,

G. D. L. P.; Dalla Rosa, C.; Mossi, A. J.; Treichel, H. (2017). Ultrasonics Sonochemistry Ultrasound-assisted hydrolysis of waste cooking oil homemade catalyzed by lipases. *Ultrasonics Sonochemistry*, *35*, 313-318.

Okoro, O. V.; Sun, Z.; Birch, J. (2018). *Lipases for Biofuel Production.* New Zealand.

Ooi, C. W.; Tey, B. T.; Hii, S. L.; Kamal, S. M. M.; Lan, J. C. W.; Ariff, A.; Ling, T. C. (2009). Purification of lipase derived from *Burkholderia pseudomallei* with alcohol/salt-based aqueous two-phase systems. *Process Biochemistry*, *44*, 1083–1087.

Ooi, C. W.; Tey, B. T.; Hii, S. L.; Kamal, S. M. M.; Lan, J. C. W.; Ariff, A.; Ling, T. C. (2011). Extractive fermentation using aqueous two-phase systems for integrated production and purification of extracellular lipase derived from *Burkholderia pseudomallei*. *Process Biochemistry*, *46*, 68-73.

Panas, P.; Lopes, C.; Cerri, M. O.; Ventura, S. P. M.; Santos-Ebinuma, V. C.; Pereira, J. F. B. (2017). Purification of clavulanic acid produced by *Streptomyces clavuligerus* via submerged fermentation using polyethylene glycol/ cholinium chloride aqueous two-phase systems. *Fluid phase equilibria*, *450*, 42-50.

Pena-Pereira, F.; Lavilla, I.; Bendicho, C. (2010). Liquid-phase microextraction techniques within the framework of green chemistry. *Trends in Analytical Chemistry*, *29*, 617–628.

Pereira, M. M.; Pedro, S. M.; Quental, M. V.; Lima, Á. S.; Coutinho, J. A. P.; Freire, M. G. (2015). Enhanced extraction of bovine serum albumin with aqueous biphasic systems of phosphonium- and ammonium-based ionic liquids. *Journal of Biotechnology*, *206*, 17-25.

Ping, L.; Yuan, X.; Zhang, M.; Chai, Y.; Shan, S. (2018). Improvement of extracellular lipase production by a newly isolated start Yarrowia mutant and its application in the biosynthesis of L-ascorbyl palmitate. *International Journal of Biological macromolecules*, *106*, 302-311.

Purwanto, M. G. M. (2016). The Role and Efficiency of Ammonium Sulphate Precipitation in Purification Process of Papain Crude Extract. *Procedia Chemistry, 18*, 127-131.

Raja, S.; Murty, V. R.; Thivaharan, V.; Rajasekar, V.; Ramesh, V. (2011). Aqueous Two Phase Systems for the Recovery of Biomolecules – A Review. *Science and Technology, 1,* 7-16.

Raja, S.; Murty, V. R.; Thivaharan, V.; Rajasekar, V.; Ramesh, V. (2012). Aqueous Two Phase Systems for the Recovery of Biomolecules – A Review. *Science and Technology, 1,* 7-16.

Ramakrishnan, V.; Goveas, L. C.; Suralikerimath, N.; Jampani, C.; Halami, P. M.; Narayan, B. (2016). Extraction and purification of lipase from *Enterococcus faecium* MTCC5695 by PEG/phosphate aqueous two-phase system (ATPS) and its biochemical characterization. *Biocatalysis and Agricultural Biotechnology, 6,* 19-27.

Rao, M. A.; Scelza, R.; Acevedo F.; Diez, M. C.; Gianfreda, L. (2014). Enzymes as useful tools for environmental purposes. *Chemosphere, 107,* 145-162.

Rao, M. B.; Tanksale, A. M.; Ghatge, M. S.; Deshpande, V. V. (1998). Molecular and Biotechnological Aspects of Microbial Proteases. *Microbiology and Molecular Biology Reviews, 62,* 597-635.

Sánchez-González, M., Alagón, A., Rodríguez-Sotrés, R., López-Munguía, A. (1999). Proteolytic processing of dextransucrase of *Leuconostoc mesenteroides. FEMS Microbiology Letters, 181,* 25–30.

Sankaran, R.; Show, P. L.; Lee, S. Y.; Yap, Y. J.; Ling, T. C. (2018a). Integration process of fermentation and liquid biphasic flotation for lipase separation from *Burkholderia cepacia. Bioresource Technology, 250,* 306-316.

Sankaran, R.; Show, P. L.;Yee Jiun Yap, Y. J.; Hon Loong Lam, H. L.; Ling, T. C.; Pan, G.; Yang, T. C. K. (2018b). Sustainable approach in recycling of phase components of large scale aqueous two-phase flotation for lipase recovery. *Journal of Cleaner Production, 184,* 938-948.

Saravanan, S.; Rao, J. R.; Nair, B. U.; Ramasami, T. (2008). Aqueous two-phase poly(ethylene glycol)–poly(acrylic acid) system for protein partitioning: Influence of molecular weight, pH and temperature. *Process Biochemistry, 43,* 905-911.

Saxena, R. K.; Sheoran, A.; Giri, B.; Davidson, W. S. (2003). Purification strategies for microbial lipases. *Journal of microbiological methods*, *52*, 1-18.

Sethi, B. K.; Nanda, P. K.; Sahoo, S. (2016). Characterization of biotechnologically relevant extracellular lipase produced by *Aspergillus terreus* NCFT 4269. 10. *Brazilian Journal of Microbiology*, *47*, n.1, 143-149.

Shao, M.; Zhang, X.; Li, N.; Shi, J.; Zhang, H.; Wang, Z.; Zhang, H.; Yu, A.; Yu, Y. (2014). Ionic liquid-based aqueous two-phase system extraction of sulfonamides in milk. *Journal of Chromatography B*, *961*, 5-12.

Show, P.; Ling, T.; Lan, J. C.; Tey, B.; Ramanan, R. N.; Yong, S.; Ooi, C. (2015). Review of microbial lipase purification using aqueous two-phase systems. *Current Organic Chemistry*, *9*, 19-29.

Shukla, S. K.; Pandey, S.; Pandey, S. (2017). Applications of ionic liquids in biphasic separation: Aqueous biphasic systems and liquid-liquid equilibria. *Journal of Chromatography A*, https://doi.org/10.1016/j.chroma.2017.10.019

Silva, M. E.; Franco, T. T. (1999). Liquid-liquid extraction of biomolecules in downstream processing - a review paper. *Brazilian Journal of Chemical Engineering*, *17*, 1-17.

Silva, O. S.; Gomes, M. H. G.; Oliveira, R. L.; Porto, A. L. F.; Converti, A.; Porto, T. S. (2017). Partitioning and Extraction Protease from *Aspergillus tamarii* URM4634 Using PEG- Citrate Aqueous Two-Phase Systems. *Biocatalysis and Agricultural Biotechnology, 9*, 168-173.

Silvério, S. C.; Rodríguez, O.; Tavares, A. P. M.; Teixeira, J. A.; Macedo, E. A. (2013). Laccase recovery with aqueous two-phase systems: Enzyme partitioning and stability. *Journal of Molecular Catalysis B: Enzymatic, 87*, 37-43.

Sousa, R. de C.; Pereira, M. M. Freire, M. G.; Coutinho, J. A. P. (2018). Evaluation of the effect of ionic liquids as adjuvants in polimer-based aqueous biphasic systems using biomolecules as molecular probes. *Separation and Purification Technology, 196*, 244-253.

Souza, R. L.; Lima, R. A.; Coutinho, J. A. P.; Soares, C. M. F.; Lima, A. S. (2015). Aqueous two-phase systems based on cholinium salts and tetrahydrofuran and their use for lipase purification. *Separation and Purification Technology, 155*, 118–126.

Targovnik, A. M.; Cascone, O.; Miranda, M. V. (2012). Extractive purification of recombinant peroxidase isozyme c from insect larvae in aqueous two-phase systems. *Separation and Purification Technology, 98*, 199-205.

Thakur, S. (2012). Lipases, its sources, properties and Applications: A Review. *International Journal of Scientific & Engineering Research, 3*, 1-29.

Treichel, H.; de Oliveira, D.; Mazutti, M. A.; Di Luccio, M.; Oliveira, V.J. (2010). A review on microbial lipases Production. *Food Bioprocess Technol, 3*, 182-196.

Tongul, B.; Kavakcıoglu, B.; Tarhan, L. (2016). Partitioning and purification of menadione induced NAD(P)H oxidase from *Phanerochaete chrysosporium* in aqueous two-phase systems. *Separation and Purification Technology, 163*, 275-281.

Tzeng, Y. P.; Shen, C. W.; Yu, T. (2008). Liquid–liquid extraction of lysozyme using a dye-modified ionic liquid. *Journal of Chromatography, 1193*, 1-6.

Ventura, S. P. M.; De Barros, R. L. R.; Barbosa, J. M. P.; Soares, C. M. F.; Lima, A. S.; Coutinho, J. A. P. (2012). Production and purification of an extracellular lipolytic enzyme using ionic liquid-based aqueous two-phase systems. *Green Chemistry, 14*, 734-740.

Ventura, S. P. M.; Neves, C. M. S. S.; Freire, M. G.; Marrucho, I. M.; Oliveira, J.; Coutinho, J. A. P. (2009). Evaluation of Anion Influence on the Formation and Extraction Capacity of Ionic-Liquid-Based Aqueous Biphasic Systems. *Journal of Physical Chemistry, 113*, 9304-9310.

Vijayaraghavan, P.; Raj, S. R. F.; Vincent, S. G. P. (2016). Industrial Enzymes: Recovery and Purification Challenges. *Agro-Industrial Wastes as Feedstock for Enzyme Production*, 95-110.

Villeneuve, P.; Muderhwa, J. M.; Graille, J.; Haas, M. J. (2000). Customizing lipases for biocatalysis: a survey of chemical, physical and molecular biological approaches. *Journal of molecular catalysis*, 9, 113-148.

Vobecká, L.; Romanov, A.; Slouka, Z.; Hasal, P.; Pribyl, M. (2018). Optimization of aqueous two-phase systems for the production of 6-aminopenicillanic acid in integrated microfluidic reactors-separators. *New Biotechnology*. https://doi.org/10.1016/j.nbt.2018.03.005.

Wanderley, M. C. A.; Neto, J. M. W. D.; Albuquerque, W. W. C.; Marques, D. A. V.; Lima, C., A.; Silvério, S. I. C.; Lima Filho, J. L.; Teixeira, J. A. C.; Porto, A. L. F. (2017). Purification and characterization of a collagenase from *Penicillium* sp. UCP 1286 by polyethylene glycol-phosphate aqueous two-phase system. *Protein expression and purification, 133*, 8-14.

Welton, T. (1999). Room-temperature ionic liquids. solvents for synthesis and catalysis, *Chemical Reviews*, 99, 2071–2083.

Whiteley, C. G.; Lee, D.-J. (2006). Enzyme technology and biological remediation. *Enzyme and Microbial Technology, 38*, 291-316.

Wilkes, J. S. (2004). Properties of ionic liquid solvents for catalysis. *Journal of Molecular Catalysis A: Chemical, 214*, 11-17.

Wolf-Márquez, V. E.; Martínez-Trujillo, M. A.; Osorio, G. A.; Patiño, F.; Álvarez, M. S.; Rodríguez, A.; Sanromán, M. A.; Deive, F. J. (2017). Scaling-up and ionic liquid-based extraction of pectinases from *Aspergillus flavipes* cultures. *Bioresource Technology, 225*, 326-335.

Yang, H.; Chen, L.; Zhou, C.; Yu, X.; Yagoub, A. E. A.; Ma, H. (2018). Improving the extraction of L-phenylalanine by the use of ionic liquids as adjuvants in aqueous biphasic systems. *Food Chemistry, 245*, 346-352.

Yongqiang, Z.; Tichang, S.; Qingxia, H.; Qing, G.; Tieqiang, L.; Yingchao, G.; Chunhuan, Y. (2016). A green method for extraction molybdenum (VI) from aqueous solution with aqueous two-phase system without any extractant. *Separation and Purification Technology, 169*, 151-157.

Zeng, Q.; Wang, Y.; Huang, Y.; Ding, X.; Chen, J.; Xu, K. (2014) Deep eutectic solvents as novel extraction media for protein partitioning. *Analyst, 139*, 2565- 2573.

Zhang, Y.; Sun, T.; Lu, T.; Yan, C. (2016) Extraction and separation of tungsten (VI) from aqueous media with Triton X-100-ammonium sulfate-water aqueous two-phase system without any extractant. *Journal of Chromatography A, 1474*, 40-46.

Zhang, H.; Wang, Y.; Zhou, Y.; Xu, K.; Li, N.; Wen, Q.; Yang, Q. (2017). Aqueous biphasic systems containing PEG-based deep eutectic solvents for high-performance partitioning of RNA. *Talanta, 170*, 266-274.

Zhang, H.; Wang, Y.; Zhou, Y.; Chen, J.; Wei, X.; Xu, P. (2018). Aqueous biphasic systems formed by deep eutectic solvent and new-type salts for the high-performance extraction of pigments. *Talanta, 181*, 210-216.

Zhou, Y. J.; Hu, C. L.; Wang, N.; Zhang, W.; Yu, X. (2013).Purification of porcine pancreatic lipase by aqueous two-phase systems of polyethylene glycol and potassium phosphate. *Journal of Chromatography B, 926*, 77-82.

In: Aqueous Two-Phase Systems   ISBN: 978-1-53614-241-9
Editor: V. Anatolijs Xanthopoulos   © 2018 Nova Science Publishers, Inc.

Chapter 3

USE OF AQUEOUS PEG-BASED TWO PHASE SYSTEMS FOR EXTRACTION OF HEAVY METAL IONS

Laura Bulgariu[1,] and Dumitru Bulgariu[2,3]*
[1]Department of Environmental Engineering and Management,
Faculty of Chemical Engineering and Environmental Protection,
Technical University Gheorghe Asachi of Iasi, Iasi, Romania
[2]Department of Geology and Geochemistry,
Faculty of Geography and Geology,
"Al.I.Cuza" University of Iasi, Iasi, Romania
[3]Romanian Academy, Filial of Iasi,
Branch of Geography, Iasi, Romania

ABSTRACT

Compared to the conventional solvent extraction, which use toxic, flammable and volatile organic solvents, requiring large samples and large volumes of extractants and organic solvents, and can be quite expensive, the extraction in aqueous PEG-based two-phase systems is considered a more environmental friendly and economically viable

---

* Corresponding Author Email: lbulg@tuiasi.ro.

method. This is because aqueous two-phase systems do not require the use of organic solvents, is efficient (the extraction percent can exceed 99% in well-defined experimental condition, and has low cost (most of the phase-forming components are commercially available and no so expensive). Generally, the aqueous two-phase systems are obtained by mixing an aqueous solution of certain water-soluble organic polymer (most frequently used is polyethylene glycol, PEG) with certain inorganic salts, in specific concentration. The obtained extraction systems are composed by two aqueous immiscible phases, the top one – rich in PEG, with the same role as organic phases from traditional extraction systems, and a bottom phase – salt-enriched. The extraction of metal ions in such aqueous two-phase systems depends on (i) the formed aqueous two-phase system characteristics and (ii) the properties of formed metallic species in extraction system. If the optimum characteristics of the aqueous two-phase system can be obtained via suitable selection of phase forming components, the second condition required the utilization of some extracting agents which will provide the selectivity of extraction process. In this chapter, are discussed the most important experimental parameters that affect the efficiency of metal ions extraction in aqueous PEG-based two-phase systems, to provide a starting point in the design of a suitable system for the extraction of metal ions.

Keywords: Aqueous two-phase system, metal ions extraction, polyethylene glycol, experimental parameters

INTRODUCTION

Solvent extraction is a well-known method that can be used for the removal and recovery of metal ions from aqueous solution, by adding organic solvents. The main advantages of solvent extraction, such as: rapid kinetics, high adaptability to a large number of metal ions, the possibility of quantitative recycling of organic solvents, etc., have determined this method to have numerous applications, including at industrial scale (Perescu, 1985; Sarghie, 1995).

But in the context of today's environmental policies (Sheldon, 2017), solvent extraction, in its traditional form, has been included in the category of harmful methods that are aggressive form the environment and harmful for human health. This because:

- most of organic solvents used in solvent extraction are toxic, flammable and volatile, and therefore their use required supplementary precautions,

- most of solvent extraction procedures required large volumes of organic solvents,

- numerous extracting agents used in the extraction of metal ions are expensive and are not stable during of time, and therefore the costs of solvent extraction procedures can be sometimes very high,

- the experimental conditions required by the solvent extraction procedure should be strictly respected, otherwise the efficiency of metal ions removal can drastically decrease.

In order to overcome all these important drawbacks of solvent extraction, alternative methods of extraction have been developed in recent years, for the separation and recovery of metal ions from aqueous solution. One of these is the extraction of metal ions in the aqueous two-phase systems, which compared with the traditional solvent extraction systems are considered to be environmentally safe (Shibukawa et al., 2001; Yoshikuni et al., 2005; da Silva et al., 2006).

The aqueous two-phase systems are formed when is mixing an aqueous solution of certain water-soluble organic polymer (such as polyethylene glycol (PEG), which is most widely used) with another organic polymer (such as dextran) (Zaslavsky, 1995) or certain inorganic salt aqueous solution ((NH_4)$_2SO_4$, Na_2SO_4, $MgSO_4$, $NaNO_3$, K_2HPO_4, Na_2CO_3, etc.), in specific concentrations (Ananthapadmanabhan and Goddar, 1987; Eiteman and Gainer, 1990; Hammer et al., 1994; da Silva et al., 1997; Huddleston, et al., 1999; Shibukawa et al., 2000; Graber and Taboada, 2000; Li et al., 2004; Foroutan and Khomami, 2008). Such extraction systems are composed by two aqueous phases, which are immiscible, a superior one – rich in organic polymer and that has the same role as organic phases from traditional solvent organic systems, and an inferior one – rich in the other organic polymer or in inorganic salt. A schematic representation of aqueous two-phase systems is illustrated in Figure 1.

                    Laura Bulgariu and Dumitru Bulgariu

Compared to the traditions solvent extraction systems, the aqueous two-phase systems are considered efficient, low-cost (because the required components are cheap and commercially available) and environmental friendly (because the main component of each phase is water) (Anathapadmanabhan and Goddard, 1987; Hatti-Kaul, 2001; Rodrigues et al., 2008). In addition, because organic solvents are not used in the formation of aqueous two-phase systems, extraction in such systems is considered to be in accordance with the principles of "green chemistry" and these extraction systems are potentially non-toxic, non-flammable and non-volatile (Anathapadmanabhan and Goddard, 1987; Sheldon, 2017). The possible use of an extraction system without volatile organic solvents opens up a new approach for many industrial processes, including those for extraction of metal ions, because it significantly reduces the negative impacts on the environment.

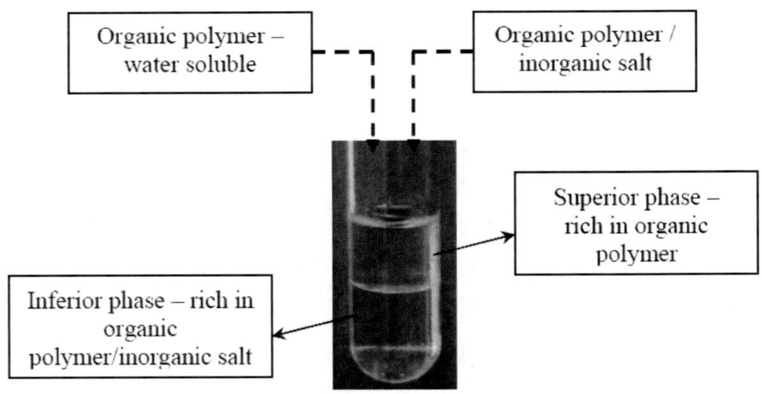

Figure 1. Schematic representation of aqueous PEG-based two-phase system.

The extraction of metal ions in such aqueous two-phase systems is influenced mainly by two categories of factors (Zaslavsky, 1986; Visser et al., 2000; Cheng et al., 2005; Bulgariu and Bulgariu, 2008a; Yu et al., 2012):

- *the characteristics of aqueous two-phase system used for extraction* – which are determined by the molecular mass and concentration of organic polymer, nature and concentration of inorganic salt, pH, temperature, presence of inert species, etc.;

- *the characteristics and properties of metallic species from extraction system* – which are related to the their chemical stability, hydration degree, ion charge, etc.

If the factors in the first category affect the physical characteristics of aqueous two-phase systems (stability of extraction systems, time of phases separation, quantity of water from superior phase, etc.), the factors in the second category determine the nature of extracted metallic species, and both categories of factors play a significant role in determining the efficiency of metal ions extraction in such systems.

In this chapter were examined the most important parameters affecting the efficiency of metal ions extraction in aqueous two-phase systems to provide useful information for the design of appropriate aqueous two-phase systems that can be used for the efficient extraction of a given metal ion in specific experimental conditions.

CLASSIFICATION OF AQUEOUS TWO-PHASE EXTRACTION SYSTEMS

The main criteria that can be used for the classification of the aqueous two-phase systems are the following:

- *as a function of the nature of phase forming components*:
  - systems consisting of two organic polymers;
  - systems consisting of an organic polymer and a common inorganic salt;
  - systems consisting of an organic salt with high molecular mass and a common inorganic salt;
  - systems consisting of a liquid crystal and a common inorganic salt.

- *as a function of the nature of extracting agent*:
    - extraction systems without extracting agents;
    - extraction systems with organic extracting agents;
    - extraction systems with inorganic extracting agents.
- *as a function of extraction mechanism*:
    - physical extraction of metallic species;
    - metal chelates extraction;
    - ionic associations extraction.
- *as a function of experimental procedure*:
    - batch extractions systems;
    - continuous extraction systems: in echi-current or in contra-current.

From the multitude of water-soluble polymers, polyethylene glycol (PEG) was most commonly used to obtain these extraction systems because: it is not toxic, flammable or volatile, stable, biodegradable, and quite cheap, thus eliminating a large part from problems related to environmental pollution, problems that occur frequently when using traditional extraction systems due to the presence of organic compounds.

Thus, polyethylene glycol (PEG) in combination with a hydrophilic organic polymer or an appropriate inorganic salt (Table 1), allows the separation of two water-immiscible phases: a superior one PEG rich, and an inferior one - with a high content in hydrophilic organic polymer or inorganic salt (as is presented in Figure 1). Depending on the properties of the metallic species and the selected experimental conditions, these can be quantitatively distributed in one of the phases of the aqueous two-phase extraction system, thus allowing the extraction process to be accomplished.

The aqueous two-phase aqueous systems composed from PEG and a hydrophilic organic polymer were initially used in protein separation processes (Huddleston et al., 1999; Mobalegholeslam et al., 2018), but further studies have shown that these systems have a relatively short lifetime (at most one month), and in the case of metal ions much more efficient systems are PEG and an inorganic salt.

Table 1. Examples of inorganic salts and hydrophilic organic polymer used to obtain the aqueous PEG-based two-phase systems (Huddleston et al., 1999)

Inorganic salts			
with mono-valent anions	with divalent anions	with trivalent anions	with tetravalent anions
NaOH KOH NaF sodium formiate	Na_2SO_4 $(NH_4)_2SO_4$ Na_2CO_3 K_2CO_3 Sodium succinate Sodium tartrate	Na_3PO_4 K_3PO_4 Sodium citrate	Na_4SiO_4
Hydrophilic organic polimers			
Polyvynil alcohol Poly(vynil)- pyralidone Dextran Ficoll			

The use of aqueous two-phase systems formed from PEG and inorganic salt in the extraction of metal ions has an extraordinary potential due to its durability, non-flammability and low cost. The phases of these systems can be easily separated by sedimentation or centrifugation, and can be used in a wide range of experimental conditions (metal ions concentrations, pH, temperature, etc.).

FORMATION OF AQUEOUS PEG-BASED TWO-PHASE SYSTEMS

In the aqueous PEG-based two-phase systems, the two aqueous phases are formed from a single homogeneous phase by the addition of one or more components, and phase separation occurs when there is structural incompatibility between the components of the mixture (Huddleston et al., 1999). Although incompletely elucidated until now, the formation and separation of phases in the aqueous PEG-inorganic salt two-phase systems appears to be based on the hydration competition between the polymer and the inorganic salt, which results in an increase in dehydration of the polymer chains and phase separation. Because in these systems the main component is water (70-80%) (Ananthapadmanabhan and Goddar, 1987), it can be said that a first condition for obtaining the two-phase aqueous systems is the high water affinity of the phase-forming components.

In the case of PEG, ethylene oxide units are strongly hydrated with 2-3 molecules of water/unit (Ananthapadmanabhan and Goddar, 1987), which explains its high solubility in water. On the other hand, the ionic species of the inorganic salt present in the solution are also hydrated, their degree of hydration depends mainly on their nature and concentration. Therefore, when the two aqueous solutions (PEG and inorganic salt solution) are brought into contact, the inorganic salt ions tend to hydrate properly, ordering as many molecules of water as possible, which causes dehydration of the polymer and the formation of the two phases.

Therefore, the formation of two-phase aqueous systems clearly involves the mutual exclusion of PEG and inorganic salt, as well as their high affinity for water. It is possible that even in the homogeneous system inorganic ions are excluded from the region near the surface of the polymer solution. With the increase in the concentration of the polymer or inorganic salt, the area of exclusion increases, so the system can reach the state where, for entropic reasons, the formation of the two phases becomes favourable (Ananthapadmanabhan and Goddar, 1987). The exclusion of inorganic ions from the polymer-water interface can take place for several reasons, namely:

- due to hydration - close to the interface, the polymer may not be accessible for inorganic salt ions;
- due to the repulsive interactions between the inorganic salt anions and the partial negative charges of the etheric oxygen (this is evidenced by the binding of multivalent polyether metal cations);

- due to their small superficial tension - which gives them a hydrophobic character.

Thus, the conditions in which the separation of the two aqueous phases occurs in such systems correspond to a considerable reduction in the contact between water polymer molecules in the PEG-rich phase, and this process is enthalphically unfavourable, but entropically favourable (Mobalegholeslam et al., 2018).

The separation of the two aqueous phases is a general phenomenon that arises from the incompatibility between the polymer solution and the inorganic salt solution when the components that make up the system are present in quantities large enough to create its heterogeneity. If, in the case of PEG, due to its high water solubility, the increase in its concentration does not significantly affect the equilibrium that takes place in the extraction process, in the case of inorganic salt this can lead to some secondary processes (such as solubility of the inorganic salt not to increase concentration above a certain value, modification of extraction performances due to saline effect, etc.). The proper choice of the phase-forming components (PEG and inorganic salt) and their optimal concentrations ensures the formation of a stable two-phase system with reasonable phase separation characteristics (phase separation time, distinct interface, etc.) that can be used in extraction processes of metal ions.

SELECTION OF PHASE FORMING COMPONENTS OF AQUEOUS TWO-PHASE SYSTEMS

The efficiency of a particular aqueous PEG-based two-phase system in extraction processes is mainly determined by the nature and concentration of phase forming components: PEG and inorganic salt. The proper selection of the phase forming components is very important in the design of such extraction systems as they influence the stability of the aqueous two-phase system and the separation characteristics of the two phases, which play an important in metal ions extraction processes.

Under these conditions, in the selection of phase-forming components to obtain a suitable aqueous two-phases system which can be used in the extraction of metal ions must be considered the following aspects:

- *molecular weight and concentration of PEG solution* – which is the component that will generate the superior phase of extraction system;
- *nature and concentration of inorganic salt solution* – which is the component that will generate the inferior phase of extraction system;
- *initial pH of inorganic salt solution* – which will influence the water content of the two phases of aqueous two-phase system and the phases ratio;
- *temperature* – which influence the separation characteristics of the two phases of the aqueous two-phase system.

Influence of Molecular Weight and Concentration of PEG Solution

As mentioned above, polyethylene glycol (PEG) is the most widely used organic polymer in the preparation of aqueous two-phase systems, and its advantages are mainly related to high solubility in water, commercial availability and low cost.

Generally, in the obtaining of the aqueous two-phase systems, the PEG solutions should be in the dilute solution region where the polymer molecules are hydrated, accessible and the interactions between polymer chains can be neglected. But, the dilute solution region depends both by molecular weight and concentration of PEG (Abbott et al., 1991). The experimental studies (Cabezas, 1996; Graber et al., 2001) have shown that the aqueous two-phase systems are obtained at a lower PEG concentration as its molecular weight is higher. Generally, the increase in the molecular weight of PEG causes the decrease of terminal hydroxyl groups, which leads to the formation of a lower number of hydrogen bonds between the polymer chains (Lei et al., 1990; Graber et al., 2001). In consequence, even if the incompatibility between the solutions of phase forming components (the separation of the two aqueous immiscible phases) is obtained at lower PEG concentrations as its molecular weight is higher, the use of PEG with high molecular weight determine the formation of high hydrophobic PEG-rich phases, which have a lower water contents, and which cannot be extracted the hydrated metal ions.

Under these conditions, it is recommended that for the preparation of aqueous PEG-based two-phase systems with applications in the extraction of metal ions, to be used PEG with low molecular weight, which allows the formation of PEG-rich phases with a relative high water content, so that the extraction of metal ions to be effective. In addition, since PEG is readily soluble in water (Graber et al., 2001), the concentration of PEG solutions can be varied in a relatively large range to better control the characteristics of the formed aqueous PEG-based two-phase system.

Starting from these observations, many studies from literature use PEG with molecular weight between 1500 – 4000 g/mol for the preparation of the aqueous PEG-based two-phase systems, and a concentration of PEG solution between 25 – 40%. The PEG-rich phases of such aqueous two-phase systems have relative high water content (Bulgariu et al., 2007) and can be successfully used for the extraction of metal ions.

Influence of Nature and Concentration of Inorganic Salt Solution

The inorganic salt is the second component required for the preparation of the aqueous PEG-based two-phase systems. In general terms, the ability of an inorganic salt to form two aqueous phases in contact with a PEG solution depends on the strength of the salting-out effect of the salt and implicitly on its concentration in the system. Since, for the use of these systems in extraction processes, it is important that the concentration of inorganic salt to be as small as possible (to minimize some of the secondary processes that may occur), the inorganic salt must exert a strong salting-out effect on the solution by PEG.

The salting-out effect of an inorganic salt is directly determined by its ability to bind the water molecules in the system (Shibukawa et al., 2000), which leads to a decrease in the concentration of free water molecules, and allows separation of the phases. From practical point of view, the evaluation of the strength of the salting-out effect of a given inorganic salt can be done, at least qualitatively, using the following parameters:

- *valence of component ions* - it was experimentally proved (Ananthapadmanabhan and Goddar, 1987) that the higher is the valence of inorganic salt anion, the greater is its salting-out effect, and therefore the minimum concentration required for the formation of the aqueous two-phase system is lower. Based on this correlation it can be noted that the efficiency of inorganic salt anions to form the aqueous two-phase systems follows the order: $PO_4^{3-} > CO_3^{2-} > SO_4^{2-} > HO^-$ (Zafarani-Moatlar and Sadeghi, 2001; Taboa et al., 2001). Unfortunately, this direct correlation between the valence of ions and the salting-out effect of a particular inorganic salt is not observed and in case of salt cations. The experimental studies (Salabat, 2001) have shown that Na^+ ions are more effective than Mg^{2+} ions, in the formation of the aqueous two-phase systems.
- *lyotropic number of the salt anion* (or its position in the Hofmeister series) - is directly correlated with the existence of a low hydration degree (Rogers et al., 1996). Thus, inorganic ions with a relatively low lyotropic number can form aqueous PEG-based two-phase systems, while ions whose lyotropic number has a high value do not form such systems (Table 2).
- *free Gibbs hydration energy* (ΔG_{hydr}) – as the value of free Gibbs hydration energy of inorganic salt is higher and negative, the higher is its salting-out effect exerted on PEG. Thus, the phases separation in the aqueous PEG-based two-phase systems is

strongly promoted by inorganic anions having high and negative free Gibbs hydration energy, which causes an "ordering" of water molecules, which creates an incompatibility between the two phases (Table 3).

Table 2. The values of lyotropic number for some inorganic anions (Ananthapadmanabhan and Goddar, 1987; Rogers et al., 1996)

Anion	Lyotropic number	Phase forming component	Anion	Lyotropic number	Phase forming component
SO_4^{2-}	2.0	yes	Cl^-	10.0	no
PO_4^{3-}	3.2	yes	Br^-	11.3	no
F^-	4.8	yes	I^-	12.5	no
HO^-	5.8	yes	SCN^-	13.3	no

Table 3. The values of ΔG_{hydr} for several inorganic anions and cations used to obtain the aqueous PEG-based two-phase system (Rogers et al., 1996; Graber et al., 2000)

Inorganic ion	ΔG_{hydr}, kJ/mol	Inorganic ion	ΔG_{hydr}, kJ/mol
PO_4^{3-}	-2773	Br^-	-250
SO_4^{2-}	-1090	I^-	-220
CO_3^{2-}	-1300	Na^+	-385
HO^-	-345	Mg^{2+}	-1830
Cl^-	-270	NH_4^+	-285

Although the influence of cations on the phase formation is generally much lower than that of anions, it can be noted that the higher is the hydration of the inorganic ions, the lower is the concentration of the inorganic salt required for the formation of the aqueous PEG-based two-phase aqueous systems. Compliance with these criteria ensures the choice of the appropriate inorganic salt to obtain an effective aqueous PEG-based two-phase extraction system.

Beside the hydration degree, another aspect which should be considered in the selection of suitable inorganic salt for the obtaining of aqueous PEG-based two phase systems with applications in metal ions extraction is the possibility of its involvement in the secondary equilibriums. Thus, even if phosphates and carbonates have a high hydration degree and a high salting-out effect (Marcus, 1991; Li et al., 2004), and are therefore the first recommended in obtaining of the aqueous two-phase systems, their selection is not appropriate when such systems are to be used for the extraction of metal ions. This is because the numerous metal ions can react with the phase-forming inorganic anions and form low-soluble phosphates or carbonates which affect the efficiency of their extraction (Bulgariu and Bulgariu, 2008a).

Therefore, it is more appropriate to use sulphate salts (Na_2SO_4 or $(NH_4)_2SO_4$) which also have a high hydration degree and a high salting-out effect on PEG, but do not influence the extraction of metals ions, because most of metals sulphates are slightly soluble. There is, however, an exception, namely in case of Pb(II) ions extraction, when the use of sulphate as anion-forming phase is not adequate due to the precipitation equilibrium which occur in the salt-rich phase (Bulgariu and Bulgariu, 2009a). In this case it is preferable to replace the sulphate with nitrate ($NaNO_3$) to obtain the maximum extraction efficiency of Pb(II) ions in aqueous PEG-based two-phase systems.

Influence of Inorganic Salt Solution pH

In the formation of the aqueous PEG-based two-phase systems, the pH on inorganic salt solution has a double influence (Bulgariu and Bulgariu, 2008a):

- affects the hydrophobicity of PEG-rich phases of extraction systems,
- affects the speciation and solubility of metal ions,

even if after the phases separation, the pH of the two phases remains practically unchanged from their initial values, and the difference in pH between the two phases is large.

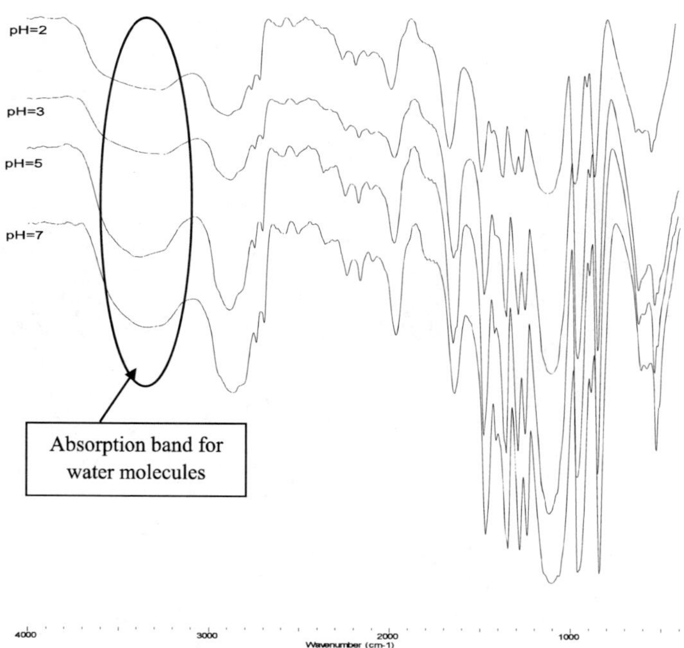

Figure 2. FTIR spectra recorded for PEG-rich phases obtained at different inorganic salt solution pH.

The influence of inorganic salt solution pH on the hydrophobicity of PEG-rich phases of aqueous two-phase systems is determined by the presence of the protons, which have a high hydration degree ($\Delta G_{hydr.}$= -1050 kJ/mol) (Marcus, 1991), and even if aren't involved in the extraction process, order a larger number of water molecules around them, increasing the salting-out effect of the inorganic salt. In consequence, the decrease of inorganic salt solution pH determines the increase of the hydrophobicity of the formed PEG-rich phases (Figure 2), which also influences the extraction efficiency of metal ions.

The variation of the extraction efficiency of various metals ions as a function of inorganic salt solution pH does not depend on the nature of inorganic salt used in the preparation of the aqueous two-phase system (Figure 3), and therefore the presence of some secondary equilibriums can be excluded.

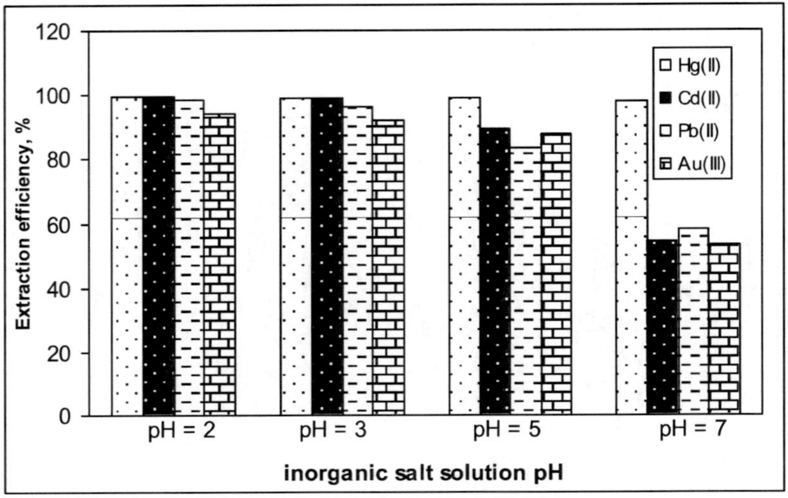

Figure 3. Influence of inorganic salt solution pH on the extraction efficiency of some metal ions in various aqueous PEG-based two-phase systems. (Experimental conditions: Hg(II), Cd(II): PEG(40%) – $(NH_4)_2SO_4$ (40%) – I^- (0.06 mol/L) (Bulgariu and Bulgariu, 2006; Bulgariu and Bulgariu, 2008b); Pb(II): PEG(40%) – $NaNO_3$ (40%) – I^- (0.06 mol/L) (Bulgariu and Bulgariu, 2009b); Au(III): PEG(40%) – $(NH_4)_2SO_4$ (40%) – Cl^- (0.06 mol/L) (Ghercă et al., 2018)).

However, all these observations are applicable only if the metal ions exist in the phases of extraction system as free species (M^{n+}). If the speciation of metal ions is changing over certain pH values, the acidity of both solutions of phase-forming components (PEG and inorganic salt) should be increased, although in this case the hydrophobicity of obtained PEG-rich phases is lower (Bulgariu and Bulgariu, 2009b).

INFLUENCE OF TEMPERATURE

As general rule, increasing of temperature increases the dispersion of the phases and accelerates their separation (Mishima et al., 1995; Zafarani-

Moatlar and Sadeghi, 2001), which will allow the obtaining of the aqueous two-phase system with a high stability over time (Table 4).

But, from the point of view of extraction of metal ions in two-phase systems, the influence of temperature is often neglected, because:

- the significant increase in temperature (with 20-30°C) does not lead to a significant improvement in the efficiency of metal ion extraction, but extraction process becomes much higher;
- the increase of temperature cannot be made over 70-80°C, due to the possibility of fragmentation of polymer chains, which will change the composition of the aqueous two-phase system.

Table 4. Influence of temperature on the formation of aqueous PEG(6000) – K$_2$HPO$_4$ two-phase system (Mishima et al., 1995)

Temperature, K	c_{salt}^{min}, (w/w) %	c_{PEG}^{min}, (w/w) %	System stability
288.15	7.38	11.87	0.3303
308.15	6.35	10.53	0.5228
318.15	5.86	9.80	0.6384

c_{salt}^{min} and c_{PEG}^{min} are the minimum concentrations of the phase forming components to obtain the two phases.

Based on the observations presented in this paragraph, it can be said that in designing of an aqueous PEG-based two-phase system for the extraction of metal ions, the following aspects should be considered:

- the main factors responsible for the incompatibility between the PEG solution and the inorganic salt solution in aqueous medium are: hydration degree and salting-out effect of inorganic salt on PEG;
- the higher is the molecular weight of PEG, the concentration of the phase-forming components required to obtain the two aqueous phases will be lower;

- the inorganic salt with low lyotropic number have a pronounced tendency to order water molecules, and are suitable to form two-phase aqueous systems in combination with PEG;
- the higher is the salting-out effect of the inorganic salt, the lower of salt concentration is required to for the two phases;
- the decrease of inorganic salt solution pH determines the increase of the hydrophobicity of the formed PEG-rich phases, and this also influences the extraction efficiency of metal ions;
- the increase of temperature increases the stability of the aqueous two-phase system, but this effect has a little influence on the extraction of metal ions.

SELECTION OF EXTRACTING AGENTS FOR METAL IONS EXTRACTION

The extraction of metal ions in aqueous PEG-based two-phase systems is generally controlled by three main factors (Rogers et al., 1996; Bulgariu and Bulgariu, 2008a):

- physico-chemical characteristics of the aqueous two-phase systems;
- physico-chemical characteristics of metal ions (ionic size, electric charge, etc.);
- type and concentration of extracting agent.

The metal ions interact strongly with water molecules (due to their high and negative free Gibbs energy (Rogers et al., 1996) and tend to remain in the salt-rich phase of the extraction systems. Therefore, to extract the metal ions from the high hydrated salt-rich phase into PEG-rich phase (where the hydration possibilities are reduced), it is necessary to change their hydration. This can be done by using a suitable extracting agent that will react with metal ions and forms less hydrated species that will prefer the PEG-rich phase of the aqueous two-phase extraction systems (Bulgariu and Bulgariu, 2008a; Bulgariu and Bulgariu, 2013).

The extracting agents used for the metal ions extraction in the aqueous two-phase systems may be:

- *organic complexing agents* – which form stable and ease extractable metal chelates;
- *inorganic anions* – which may form with the metal ions the anionic complexes that can be extracted into PEG-rich phase through a similar mechanism to the extraction of ionic pairs with ethers or ketones.

There are also metallic species that can be extracted into PEG-rich phase without the addition of extracting agent. An example is the extraction of TcO_4^- ions, which have an advanced degree of covalence of the molecules, which makes these species not to accommodate the ordered environment from the salt-rich phase, and to pass quantitatively into the PEG-rich phase of extraction system (Rogers et al., 1996; Bulgariu and Bulgariu, 2008a).

However, regardless of the nature of the extraction agent, the extraction of metal ions in aqueous PEG-based two-phase system involves a succession of elementary equilibriums, schematically illustrated in Figure 4.

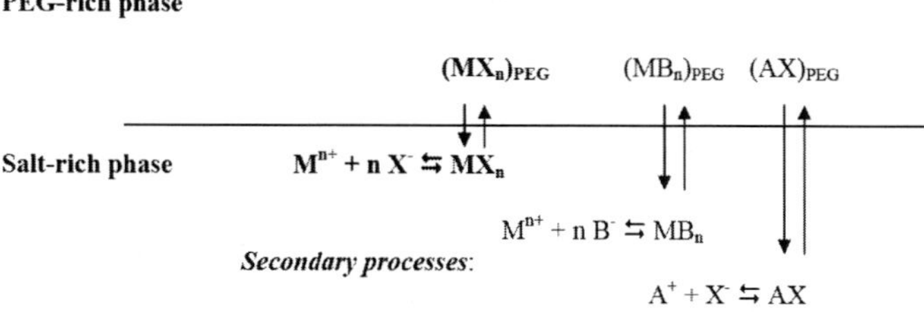

Figure 4. Schematic representation of the metal ions extraction in aqueous PEG-based two-phase systems (M^{n+} is the metal ion; X^- is the extracting agent (organic or inorganic); A^+ and B^- are the cation and anion of phase-forming inorganic salt).

The minimization of the secondary processes that occur can be done, at least theoretically, by appropriate selection of the extraction system components, as follow:

- the inorganic salt-forming phase (AB) do not form with the metal ions and extractable species in extraction system;
- the extracting agent (X^-) must be high soluble in water and to form stable complexes with metal ions.

Starting from these considerations, Roger and his collaborators (Roger et al., 1996) classified metal ion extraction in aqueous PEG-based two-phase systems into three categories:

- extraction of metal ions without an extracting agent;
- extraction of metal ions as chelates using water-soluble organic chelating extractants;
- extraction of metal ions as anionic complexes with inorganic ions.

EXTRACTION OF METAL IONS IN ABSENCE OF EXTRACTING AGENTS

The most eloquent example from this category is the extraction of TcO_4^- ions, which can be quantitatively extracted in the PEG-rich phase of an aqueous two-phase system, in the absence of extracting agents. The TcO_4^- ions have various applications in medical procedures using radioisotopes, in nuclear and metallurgical industry, and therefore numerous studies have been directed towards finding of some adequate aqueous two-phase systems for their extraction and separation (Rogers and Zhang, 1996).

The extraction of TcO_4^- ions in the aqueous PEG-based two-phase systems are due to their pronounced hydrophobic nature (advanced covalent degree) and relatively low hydration free Gibbs energy(ΔG_{hydr} = -

251 kJ/mol (Marcus, 1991)), which makes these ions more compatible with the PEG-rich phase than the salt-rich phase of extraction system.

Table 5. Distribution coefficients (D(TcO$_4^-$)) of TcO$_4^-$ ions in the aqueous PEG-based two-phase systems formed with different inorganic salts (Marcus, 1991; Rogers et al., 1996)

	Na$^+$			SO$_4^{2-}$		
Inorganic ion	SO$_4^{2-}$	CO$_3^{2-}$	CrO$_4^{2-}$	Na$^+$	NH$_4^+$	Cs$^+$
ΔG_{hydr}, kJ/mol	-1145	-1300	-1120	-385	-285	-245
D(TcO$_4^-$)	13.59	12.01	8.53	13.59	9.65	8.07

In the absence of an extracting agent to change the "hydration properties" of these anions, the proper choice of the phase-forming components (PEG and inorganic salt) is the key factor obtaining a quantitative extraction. Thus, the extraction efficiency, in this case, can be increased by:

- *choosing an inorganic salt that to exerts a strong salting-out effect of PEG.* The sulphates appear to be one of the optimal choice because of their high solubility in water and excellent salting-out effect on PEG, which allow the formation of PEG-rich phases compatible with the hydration of TcO$_4^-$ ions (Table 5).

Table 6. Distribution coefficients (D(TcO$_4^-$)) of TcO$_4^-$ ions in the aqueous PEG-based two-phase systems as a function of molecular weight of PEG (Rogers and Zhang, 1996)

Molecular weight of PEG	2000	3400	12000
D(TcO$_4^-$)	13.33	18.01	20.01

Experimental studies have shown that the distribution coefficient of TcO$_4^-$ ions increases with the increase in the concentration of the inorganic phase-forming salt, and this is due, both to the decrease in the degree of hydration of the TcO_4^- ions, and to the increase in PEG-rich phase viscosity (Rogers et al., 1996; Rogers and Zhang, 1996).

- *proper choice of molecular weight of PEG* – because it is known that the incompatibility between the two-phases of the aqueous two-phase systems is strongly influenced by the molecular weight of PEG (Cabezas, 1996; Graber et al., 2001). Therefore, an increase in the molecular weight of the PEG is expected to increase the distribution coefficient of TcO_4^- ions in the aqueous PEG-based two-phase systems (Table 6).

Practically, there is a limitation in the choice of the molecular weight of PEG for such systems, since with the increase of molecular weight of PEG determines a significant increase of the PEG-rich phase viscosity. Moreover, if the polymer used for the preparation of the aqueous two-phase system is "too" hydrophobic, the TcO_4^- ions can also be excluded and the extraction does not take place.

The quantitative extraction of TcO_4^- ions in the aqueous PEG-based two-phase systems in absence of extracting agents suggests the possibility that even other oxoanions (such ReO_4^-, WO_4^{2-} or MoO_4^{2-}) could be extracted in similar conditions. Although, no studies have been conducted to confirm the extraction behaviour of these oxoanions, the estimations based on free Gibbs hydration energy suggest that the ReO_4^- ions will be less extracted than TcO_4^- ions, while WO_4^{2-} and MoO_4^{2-} ions will preferring to remain in the salt-rich phase of extraction system (Rogers and Zhang, 1996).

EXTRACTION OF METAL IONS USING ORGANIC CHELATING EXTRACTANTS

The extraction of metal ions with organic chelates extractants is mainly used for the separation and recovery of highly hydrated metal ions, such as lanthanide ions and some transitional metal ions. Such metal ions interact strongly with the water molecules (have high and negative free Gibbs hydration energy) and therefore they will prefer to remain predominant in the salt-rich phase of the aqueous two-phase system. The use of an organic chelating agent as extractant will determine the drastically decrease in metal ion hydration, due to the formation of a non-ionizable chelate, which can be easily extracted in the PEG-rich phase of the system.

The organic chelating agents that can be used for the extraction of metal ions in these systems must meet the following conditions (Bulgariu and Bulgariu, 2008a):

- have to be readily soluble in water;
- have to be extracted quantitatively in the PEG-rich phase of the aqueous two-phase system;
- have to form with the metal ions a stable non-ionizable chelate, with a low hydration degree; requirements that drastically reduce their number.

The organic chelating extractants frequently used in the traditional solvent extraction systems cannot be used for extraction in the aqueous PEG-based two-phase systems, mainly because of their very low solubility in aqueous solution. Only, o-phenanthroline is mentioned in literature as an extracting agent for the extraction of Cu(II) ions in PEG 3350 − $(NH_4)_2SO_4$ two-phase system, but with moderate performances (Fontana and Ricci, 2000).

More suitable for such extraction systems have been shown to be organic dyes (such as: arsenazo III, xylenol orange, alizarin, etc.) as well as some macrocyclic ligands (such as crown ethers, calixarene), which are readily soluble in water and form stable and low hydrated metal chelates with some metal ions (Table 7).

The main drawbacks in the practical utilization of the organic chelating extractants for the extraction of metal ions in the aqueous PEG-based two-phase systems are determined by the fact that these extracting agents are very expensive and their efficiency in the separation processes is moderate.

**Table 7. Examples of chelating agents used for the metal ions
extraction in the aqueous PEG-based two-phase systems**

Cheating agent	Aqueous two-phase system	Extracted metal ions	References
Arsenazo III	PEG 2000-$(NH_4)_2SO_4$	Th^{4+}, Pu^{4+}, U^{4+}	Rogers et al., 1993;
Alizarin	PEG 2000-$(NH_4)_2SO_4$	Th^{4+}, Pu^{4+}	Shkinev et al., 1985
	PEG 2000-M_2CO_3	actinide ions	
Xylenol orange	PEG 2000-$(NH_4)_2SO_4$	Pu^{4+}, UO_2^{2+}, Th^{4+}, Am^{3+}	Rogers et al., 1993;
18-crown-6 ether	PEG 2000-$(NH_4)_2SO_4$	UO_2^{2+}, Pu^{4+}, Th^{4+}, Am^{3+}	Rogers et al., 1993;
	PEG 2000-NaOH	Cs^+, UO_2^{2+}, Na^+, Ba^{2+}, Sr^{2+}, Cd^{2+}, Co^{2+}	Rogers and Bauer, 1996

EXTRACTION OF METAL IONS USING INORGANIC ANIONS AS EXTRACTING AGENTS

Another way in which the metal ions can be transformed into extractable species and extracted in the aqueous PEG-based two-phase systems is by using inorganic anions as extracting agents. Most of transitional metal ions, as well as some ions of lanthanides and actinides, which form with halide and pseudo-halide ions stable anionic complexes (most often MX_4^{2-} type) can be extracted into aqueous PEG-based two-phase system most likely by an ion association mechanism (Rogers et al., 1996; Bulgariu and Bulgariu, 2008a). In addition, the inorganic ions such as I^-, Br^-, Cl^-, SCN^- or CN^- are readily soluble in water, and according with the studies from literature (Table 8) can be quantitatively distributed in the PEG-rich phase of the aqueous two-phase system, thus fulfilling the two requirements of their use as extracting agents.

Table 8. The distribution coefficients (D_{X^-}) of some inorganic anions in the aqueous PEG(2000) – $(NH_4)_2SO_4$ two-phase system (Marcus, 1991; Rogers et al., 1996)

Anion	I^-	Br^-	Cl^-	SCN^-	CN^-
ΔG_{hydr}, kJ/mol	-220	-250	-270	-230	-260
D_{X^-}	10	3.2	1.7	5.0	2.4

The efficiency of halide and pseudo-halide ions as extracting agents in the aqueous PEG-based two-phase systems is directly correlated with their properties. Thus, it is expected that the inorganic anions that form with metal ions stable anionic complexes with low hydration degree to be efficient extracting agents in the extraction processes using aqueous PEG-based two-phase systems.

Under these conditions, it is normal to assume that in the extraction of metal ions are involved several elementary equilibriums (Table 9), which result in the formation of the most stable and less hydrated metal species, in the particular experimental conditions of extraction system.

Table 9. The main elementary equilibriums involved in the extraction of metal ions using inorganic anions extractants, in aqueous PEG-based two-phase systems (Bulgariu and Bulgariu, 2008a)

Elementary equilibrium	Equilibrium reaction
Formation of metallic species in salt-rich phase	$M^{m+} + m\,A^- \leftrightarrows MA_m^{y-}$
Formation of metal anionic complexes	$M^{n+} + n\,X^- \leftrightarrows MX_n^{x-}$
Inorganic extractant distribution	$X^- \leftrightarrows (X^-)_{PEG}$
Metal anionic complexes distribution	$MX_n^{x-} \leftrightarrows (MX_n^{x-})_{PEG}$
Distribution of other anionic metallic species	$MA_m^{y-} \leftrightarrows (MA_m^{y-})_{PEG}$

Notations: The chemical species in the PEG-rich phase are noted with "PEG" subscript, while the chemical species from the salt-rich phases are written without any subscript; A^- is the phase-forming anion; X^- is the inorganic anion extractant.

On the basis of the elementary equilibrium from Table 9, the distribution coefficient (D_M) for the metal ions extraction in the aqueous

                    Laura Bulgariu and Dumitru Bulgariu

PEG-based two-phase systems in presence of inorganic anions extractants, can be written as:

$$D_M = \frac{[MA_m^{y-}]_{PEG} + \sum[MX_n^{x-}]_{PEG}}{[MA_m^{y-}] + \sum[MX_n^{x-}]} \tag{1}$$

This equation indicates that the efficiency of metal ions extraction in the aqueous PEG-based two-phase systems in presence of inorganic anions extractants mainly depends by two factors (Bulgariu and Bulgariu, 2008a):

- *stability of the metal anionic complexes formed between metal ions and inorganic extractant (MX_n^{x-})* – because as the stability of this specie is higher, the concentration of MA_m^{y-} and $(MA_m^{y-})_{PEG}$ species are smaller and the distribution coefficients will be high;
- *hydration degree of MX_n^{x-} species* – which is directly responsible by the distribution of MX_n^{x-} species between the two phases of the extraction system and it is influenced by the characteristics of the aqueous PEG-based two-phase system.

Considering the influence of these two factors, it can be explained the experimental results obtained by us, which have shown that the same metal ion can be extracted as different species using the same extracting agent and in the same aqueous PEG-based two-phase system (Table 10).

As can be observed from Table 10, always at low pH value of inorganic salt phase-forming solution, the number of inorganic anions extractants is higher than in basic media. These results are in agreement with the previous observations which indicate that the increase of inorganic salt phase-forming solution acidity determined the decrease of water content from the PEG-rich phases of the extractions systems, which means that the hydration degree of the extracted metallic species must be also low (Bulgariu and Bulgariu, 2008a; 2013).

Table 10. Type of metallic species extracted in the aqueous PEG-based two-phase systems

Metal ion	Extractant	Acid media	Basic media	Reference
PEG(1550) – (NH$_4$)$_2$SO$_4$				
Hg(II)	I$^-$	HgI$_4^{2-}$	HgI$_3^-$	Bulgariu and Bulgariu, 2006
	Br$^-$	HgBr$_4^{2-}$	HgBr$_3^-$	
	Cl-	HgCl$_3^-$	HgCl$_3^-$	
Cd(II)	I$^-$	CdI$_4^{2-}$	CdI$_2$	Bulgariu and Bulgariu, 2008b
	Br$^-$	CdBr$_4^{2-}$	CdBr$_2$	
	Cl-	CdCl$_3^-$	CdCl$_2$	
PEG(1550) – NaNO$_3$				
Pb(II)	I$^-$	PbI$_3^-$	PbI$^+$	Bulgariu and Bulgariu, 2009a
	Br$^-$	PbBr$_3^-$	PbBr$^+$	
	Cl-	PbCl$_3^-$	PbCl$^+$	

Table 11. The use of inorganic anion extractants for the extraction of metal ions in aqueous PEG-based two-phase systems

Inorganic anion extractant	Experimental conditions	Extracted metal ions
I$^-$	0.5 – 2.0 mol/L; Acid media	Hg(II), Cd(II), Bi(III), Tl(III), Cu(I)
Br$^-$	0.5 – 2.0 mol/L; Acid media	Bi(III), Pb(II), Zn(II), Au(III), In(III), Hg(II)
Cl$^-$	0.5 – 2.0 mol/L; Acid media	Bi(III), Pb(II), Zn(II), Au(III)
SCN$^-$	1.0 – 10.0 mol/L;	Co(II), Cu(II), Zn(II), Fe(IIII), In(III)
CN$^-$	1 mol/L; pH =10.5	Au(I)

Rogers et al., 1996; Visser et al., 2000; Bulgariu and Bulgariu, 2008a; Bulgariu and Bulgariu, 2013.

In Table 11 are summarized some applications of the inorganic anion extractants for the extraction of metal ions in the aqueous PEG-based two-phase systems.

The use of inorganic anions extractants for the metal ions extraction in the aqueous PEG-based two-phase systems have several important advantages such as:

- are cheap and commercially available;
- are non-toxic, non-flammable, non-volatile, being in agreement with the principles of "green chemistry"
- are stable for a long period of time and not required further purifications, which makes their use easier;
- have high solubility in water, even at room temperature and lower Gibbs free energy of hydration;
- in well-defined experimental conditions can extract quantitatively certain metal ions (>95%);

Advantages which have determined that these extracting agents to be many times preferred in the detriment of organic chelating extractants.

CONCLUSION AND FINAL REMARKS

The aqueous two-phase extraction systems, formed from an organic water-soluble polymer and a certain inorganic salt represent a new extraction method, that don't require toxic and volatile organic solvent, and which were shown to be as effective in separation processes as traditional extraction systems. The aqueous two-phase systems are prepared by mixing an aqueous solution of certain water-soluble organic polymer (most frequently used is polyethylene glycol, PEG) with certain inorganic salt, in specific concentration, when are formed two aqueous immiscible phases, the top one – rich in PEG, with the same role as organic phases from traditional extraction systems, and a bottom phase – salt-enriched. Because, the formation of aqueous two-phase systems two-phase systems doesn't involve the use of any organic solvents, these extraction systems can be considered virtually non-toxic, non-flammable and non-volatile, and from these reasons they are strategically compatible with the principles of green chemistry.

The efficiency of an aqueous PEG-based two-phase system in extraction processes of metal ions mainly depends by the nature and concentration of phase forming components: PEG and inorganic salt.

Because the phase forming components influences the stability and characteristics of phases separation, their selection is an important step in the design of an aqueous two-phase system for metal ions extraction.

Although there are cases when metal ions can be extracted into PEG-rich phase without the addition of an extracting agent, most extraction processes occur in presence of extracting agents, which can be organic or inorganic, as a function of the nature of metal ion. However, considering the main advantages it can be said that the inorganic anion extractants are generally preferred in the detriment of organic chelating extractants, for the extraction of metal ions in the aqueous PEG-based two-phase systems.

ACKNOWLEDGMENTS

This study was elaborated with the support of grant of the Romanian National Authority for Scientific Research, CNCS – UEFISCDI, project number PN-III-P4-ID-PCE-2016-0500.

REFERENCES

Abbott, N. L., Blankschtein, D., Hatton, T. A., (1991). Protein partitioning in two-phase aqueous polymer systems. 1. Novel physical pictures and a scaling thermodynamic formulation, *Macromolecules* 24, 4334-4348.

Ananthapadmanabhan, K. P., Goddar, E. D., (1987). Aqueous Biphase Formation in Polyethylene Oxide-Inorganic Salt Systems. *Langmuir*, 3(1), 25-31.

Bulgariu, L., Bulgariu, D., (2006). Hg (II) Extraction in a PEG-Based Aqueous Two-Phase System in the Presence of Halide Ions. I. Liquid Phase Analysis. *Centr. Europ. J. Chem.* 4(2), 246-257.

Bulgariu, L., Bulgariu, D., Sârghie, I, Malutan, T., (2007) Cd(II) Extraction in PEG-based Two-Phase Aqueous Systems in the Presence of Iodide Ions. Analysis of PEG-rich Solid Phases. *Central Europ. J. Chem.* 5(1), 291-302.

Bulgariu, L., Bulgariu, D., (2008a). Extraction of Metal Ions in Aqueous Polyethylene Glycol–Inorganic Salt Two-Phase Systems in the Presence of Inorganic Extractants: Correlation between Extraction Behaviour and Stability Constants of Extracted Species. *J. Chromatogr. A* 1196-1197, 117-124.

Bulgariu, L., Bulgariu, D., (2008b). Cadmium Extraction in PEG(1550) – $(NH_4)_2SO_4$ Two-Phase Systems Using Halide Extractants. *J. Serb. Chem. Soc.* 73(3), 341-350.

Bulgariu, L., Bulgariu, D., (2009a). Lead(II) Extraction in Aqueous PEG(1550) – $(NH_4)_2SO_4$ Two-Phase Systems Using Iodide Ions as Extracting Agent. *Bull. Polytech. Inst. Iasi* 55(3), 9-18.

Bulgariu, L., Bulgariu, D., (2009b). The Utilization of Aqueous PEG(1550) – $NaNO_3$ Two-Phase System for the Efficient Extraction of Pb(II) with Iodide Extractants. *Annals West Univ. Timisoara* 18(1), 73-80.

Bulgariu, L., Bulgariu, D. (2013). Selective extraction of Hg(II), Cd(II) and Zn(II) ions from aqueous media by a green chemistry procedure using aqueous two-phase systems. *Sep. Purif. Technol.* 118, 209-216.

Cabezas, H., (1996). Theory of Phase Formation in Aqueous Two-Phase Systems. *J. Chromatogr. B* 680(1-2), 3-30.

Cheng, J., Spear, S. K., Huddleston, J. G., Rogers, R. D., (2005). Polyethylene Glycol and Solutions of Polyethylene Glycol as Green Reaction Media. *Green Chem.* 7(2), 64-82.

da Silva, L. H. M., Coimbra, J. S. R., de A Meirelles, A. J., (1997). Equilibrium Phase Behaviour of Poly(ethylene glycol) + Potassium Phosphate + Water Two-Phase Systems at Various pH and Temperatures. *J. Chem. Eng. Data* 42(2), 398-401.

da Silva, M. D. H., da Silva, L. H. M., Paggioli, F. J., Reis Coimbra, J. S., Minim, L. A., (2006). Aqueous Biphasic Systems: an Efficient Alternative for Extraction of Ions. *Quimica Nova*, 29(6), 1332-1339.

Eitheman, M. A., Gainer, J. L., (1990). Peptide Hydrophobicity and Partitioning in Poly(ethylene glycol)/ Magnesium Sulphate Aqueous Two-Phase Systems. *Biotechnol. Progr.* 6(6), 479-484.

Fontana, D., Ricci, G., (2000). Poly(ethylene glycol)-Based Aqueous Biphasic Systems: Effect of Temperature on Phase Equilibrium and on Partitioning of 1,10-phenanthroline–Copper(II) Sulphate Complex. *J. Chromatogr. B* 743(1-2), 231-234.

Foroutan, M., Khomami, M. H., (2008). Activities of Polymer, Salt and Water in Liquid–Liquid Equilibria of Polyvinylpyrrolidone and K_2HPO_4/KH_2PO_4 Buffer using the Flory–Huggins Model with Debye–Huckel Equation and the Osmotic Virial Model: Effects of pH and Temperature. *Fluid Phase Equilibria* 265(1-2), 17-24.

Gherca, M., Bulgariu, D. Mocanu, A. M., Bulgariu, L., (2018). Green chemistry extraction method for the recovery of gold ions from cyanide wastewater. *Enviornm. Eng. Manag. J.* (in press).

Graber, T. A., Taboada, M. E., (2000). Liquid-Liquid Equilibrium of the Poly(ethylene glycol) + Sodium Nitrate + Water System at 298.15 K. *J. Chem. Eng. Data* 45(2), 182-184.

Graber, T. A., Andrews, B. A., Asenjo, J. A., (2000). Model for the Partition of Metal Ions in Aqueous Two-Phase Systems. *J. Chromatogr. B*, 743(1-2), 57-64.

Graber, T. A., Taboada, M. E., Asenjo, J. A., Andrews, B. A., (2001). Influence of Molecular Weight of the Polymer on the Liquid-Liquid Equilibrium of the Poly(ethylene glycol) + $NaNO_3$ + H_2O System at 298.15 K. *J. Chem. Eng. Data* 46(3), 765-768.

Hammer, S., Phennig, A., Stumph, M., (1994). Liquid-Liquid and Vapor-Liquid Equilibria in Water + Poly(ethylene glycol) + Sodium Sulfate. *J. Chem. Eng. Data* 39(3) 409-413.

Hatti-Kaul, R., (2001). Aqueous Two-Phase Systems. A General Overview. *Molec. Biotechnol.* 19, 269-277.

Huddleston, J. G., Willauer, H. D., Griffin, S. T., Rogers, R. D., (1999). Aqueous Polymeric Solutions as Environmentally Being Liquid/Liquid Extraction Media. *Ind. Eng. Chem. Res.* 38(6), 2523-2539.

Lei, X., Diamont, A. D., Hsu, J. T., (1990). Equilibrium phase behavior of the poly(ethylene glycol)/potassium phosphate/water two-phase system at 4. degree. C *J. Chem. Eng. Data* 35, 420-423.

Li, L., He, C. Y., Li, S. H., Liu, F., Su, S., Kong, X. X., Li, N., Li, K. A., (2004). Study on PEG-$(NH_4)_2SO_4$ Aqueous Two-Phase System and Distribution Behaviour of Drugs. *Chinese J. Chem.* 22(11), 1313-1318.

Marcus, Y. (1991). Thermodynamics of Solvation of Ions. *J. Chem. Soc. Faraday Trans.* 87(18), 2995-2999.

Mishima, K., Nakatoni, K., Nomiyama, T., Matsuyama, K., Nagatoni, M., Nishikawa, H., (1995). Liquid-liquid equilibria of aqueous two-phase systems containing polyethylene glycol and dipotassium hydrogenphosphate, *Fluid Phase Equilib.*, 107, 269-276.

Mobalegholeslam, P., Moayyedi, G., Majed, A. A., (2018). A new proposed thermodynamic model for aqueous polymer solutions. *J. Molec. Liquids* 253, 53-60.

Perescu, N. E., (1985). *Solvent Extraction Chemistry and Applications* (in Romanian), R. S. R Academy Ed, Bucharest, Romania.

Rodrigues, G. D., da Silva, M. D. H., da Silva, L. H. M., Paggiolli, F. J., Minim, L. A., Coimbra, J. S. R., (2008). Liquid–liquid extraction of metal ions without use of organic solvent. *Sep. Purif. Technol.* 62(3), 687-693.

Rogers, R. D., Bond, A. H., Bauer, C. B., (1993). Metal Ion Separations in Polyethylene Glycol-Based Aqueous Biphasic Systems. *Sep. Sci. Technol.* 28(5), 1091-1126.

Rogers, R. D., Bauer, C. B., (1996). Water soluble calixarenes as possible metal ion extractants in polyethylene glycol-based aqueous biphasic systems *J. Radioanal. Nucl. Chem.* 208(1), 153-161.

Rogers, R. D., Bond, A. H., Bauer, C. B., Zhang, J., Griffin, S. T., (1996). Metal Ion Separation in Polyethylene Glycol-Based Aqueous Biphasic Systems: Correlation of Partition Behaviour with Available Thermodynamic Hydration Data. *J. Chromatogr. B* 680(1-2), 221-229.

Rogers, R. D., Zhang, J., (1996). Effects of Increasing Polymer Hydrophobicity on Distribution Ratios of TcO4- in Polyethylene/ Poly(propylene glycol)-Based Aqueous Biphasic Systems. *J. Chromatogr. B* 680(1-2), 231-236.

Salabat, A., (2001). The influence of salts on the phase composition in aqueous two-phase systems: experiments and predictions. *Fluid Phase Equilib.*, 187-188, 489-498.

Sarghie, I., (1995). *Extraction Mechanism of Metal Complexes* (in Romanian), Gheorghe Asachi Ed, ISBN 973-9178-14-6, Iaşi, Romania.

Sheldon, R. A., (2017). The *E* factor 25 years on: the rise of green chemistry and sustainability. *Green Chem.* 19, 18-43.

Shibukawa, M., Matsuura, K., Shinozuka, Y., Mizuni, S., Oguma, K., (2000). Effects of Phase Forming Cations and Anions on the Partition of Ionic Solutes in Aqueous Polyethylene Glycol − Inorganic Salt Two-Phase Systems. *Anal. Sci.* 16(10), 1039-1044.

Shkinev, V. M., Malochnikova, N. P., Zvarova, T. I., Spivakov, B. Y., Myasoedov, B. F., Zolotov, Y. A., (1985). Extraction of complexes of lanthanides and actinides with Arsenazo III in an ammonium sulfate-poly(ethylene glycol)-water two-phase system. *J. Radioanal. Nucl. Chem.*, 88, 115-123.

Taboa, M. E., Aenjo, J. A., Andrews, B. A., (2001). Liquid–liquid and liquid–liquid–solid equilibrium in Na_2CO_3–PEG–H_2O. *Fluid Phase Equilib.*, 180, 273-280.

Visser, A. E., Griffin, S. T., Ingenito, C. C., Hartman, D. H., Huddleston, J. G., Rogers, R. D., (2000). Aqueous Biphasic Systems as a Novel Environmentally Benign Separation Technology for Metal Ion Removal. *Metal Separation & Technology Beyond* 119-130.

Yoshikuni, N., Baba, T., Tsunoda, N., Oguma, K., (2005). Aqueous Two-Phase Extraction of Nickel Dimethylglyoximato Complex and its Application to Spectrophotometric Determination of Nickel in Stained Steel. *Talanta*, 66(1), 40-44.

Yu, P., Huang, K., Liu, H., Xie, K., (2012). Three-Liquid-Phase Partition Behaviors of Pt(IV), Pd(II) and Rh(III): Influences of Phase-Forming Components. *Sep. Purif. Technol.* 88, 52-60.

Zafarani-Moatlar, M. T., Sadeghi, R., (2001). Liquid–liquid equilibria of aqueous two-phase systems containing polyethylene glycol and sodium dihydrogen phosphate or disodium hydrogen phosphate: Experiment and correlation. *Fluid Phase Equilib.* 181, 95-112.

Zaslavsky, B. Y., (1995). *Aqueous two-phase partitioning: physical chemistry and bioanalytical applications*, Marcel Dekker Inc., New York.

Zaslavsky, B. Y., Bagirovl, T. O., Borovskaya, A. A., Gasanoval, G. Z., Gulaeva, N. D., Levin, V. Yu., Masimov, A. A., Mahmudov, A. U., Mestechkina, N. M., Miheeva, L. M., Osipov, N. N., Rogozhin, S. V., (1986). Aqueous Biphasic Systems Formed by Nonionic Polymers I. Effects of Inorganic Salts on Phase Separation. *Coll. Polym. Sci.* 264(12), 1066-1071.

In: Aqueous Two-Phase Systems   ISBN: 978-1-53614-241-9
Editor: V. Anatolijs Xanthopoulos   © 2018 Nova Science Publishers, Inc.

Chapter 4

APPLICATION OF AQUEOUS TWO-PHASE SYSTEMS IN MICROSTRUCTURED DEVICES

Ana Jurinjak Tušek[1],, Anita Šalić[2] and Bruno Zelić[2]*
[1]University of Zagreb, Faculty of Food Technology
and Biotechnology, Zagreb, Croatia
[2]University of Zagreb, Faculty of Chemical Engineering and
Technology, Zagreb, Croatia

ABSTRACT

Aqueous two-phase systems (ATPSs) are formed when two incompatible polymers or certain polymers and lyotropic salts are mixed. They have been recognized as superior and versatile, preparative and analytical tools for the downstream processing of biomolecules with wide application in extractions, separations, purifications and enrichments of proteins, polyphenols, viruses, enzymes and other biomolecules both in research and industry. An overview about application of ATPSs in microstructured devices is given in this work. Microstructured devices offer potential benefits due to well-defined high specific interfacial areas available for heat and mass transfer. High specific interfacial area increases transfer rate and enhances yield, selectivity and process control. The flow patterns formations and stability, mass transfer and samples

---

* Corresponding Author Email: atusek@pbf.hr.

partition coefficients using ATPSs in microstructured devices are discussed in this work. Also, the application of ATPSs in Micro Total Analysis Systems (μ-TAS) is described.

Keywords: aqueous two-phase systems, extraction, microstructured devices

AQUEOUS TWO PHASE SYSTEMS

Aqueous two phase systems (ATPSs) are liquid-liquid systems composed of two immiscible polymers (e.g., polyethylene glycol, dextran), or a polymer and a salt (e.g., ammonium sulfate, potassium phosphate) that are water soluble in a certain concentrations (Xu et al., 2003; Rodrigues de Lemos et al., 2010; Goja et al., 2013). Above critical concentrations of these components, spontaneous phase separation takes place with each of the two resulting phases enriched with respect to one of the components (Xu et al., 2003). The ATPSs contain three zones; two bulk phases, often separated by density into top and a bottom phase, and interface. Water concentration in ATPSs ranges from 65% to 90%, ensuring mild environment for the extraction (Rahimpour and Baharvand, 2009). Water concentration higher than 70% generates low interfacial tension, non-flammable, slightly toxic systems safe to the environment (Liu et al., 2012). Alternative biphasic systems can be obtained using surfactants, micellar compounds or ionic liquids (Silvério et al., 2013). Aqueous biphasic systems (ABS) have been considered as a competitive partition technique due to the short process time, low viscosity, little emulsion formation, high extraction efficiency and low energy consumption (Álvarez et al., 2012). In aqueous biphasic systems hydrophobic ionic liquids are used as a phase separator either with water or with an aqueous solution of the polymer. Ionic liquids are becoming used in aqueous biphasic systems to tune the physicochemical properties of the ABS. Using ABS with ILs, much wider hydrophilic-lipophilic range is ensured, allowing for more extensive and selective separation (Almeida et al., 2014).

ATPSs ensure concentration and purification in a single operation unit; because of this, liquid system scale up is quite simple (Sutherland et al., 2011). These systems have high selectivity and recovery of biomolecules. Due to its advantages, ATPSs have been widely used in the field of biotechnology for separation and purification of various biological products, such as proteins, amino acids, enzymes, cells, antibodies and other byproducts (Teixeira et al., 2017).

TYPES OF AQUEOUS TWO-PHASE SYSTEMS

As mentioned before, three types of ATPSs have been defined: (i) polymer and polymer, (ii) polymer and salt and (iii) biphasic systems obtained by using surfactants, micellar compounds or ionic liquids. The most studied ATPSs are based upon segregative behavior of two non-charged polymers. The phase system based on two polymers occurs from unfavorable energy of interactions when segments of one polymer contact segments of the other polymer or the polymer segments bond strongly to each other (Ratanapongleka, 2010). Different polymers can be used for preparation of APTSs like polyethylene glycol (PEG), dextran, polypropylene glycol, polyvinylpyrrolidone and hydroxypropyldextran (Silverio et al., 2012). Some examples of polymer-polymer ATPSs usage are given in Table 1.

According to Lladosa et al., (2012) polymer-salts ATPSs have some advantages over two polymers ATPSs; chemicals are cheaper and the phases have lower viscosity. Therefore, shorter times are required for phase separation. Many types of salt, such as potassium dihydrogen phosphate, potassium chloride, sodium dihydrogen phosphate, sodium carbonate, sodium citrate, magnesium sulphate and ammonium sulfate, can be used in forming ATPSs with polymers, especially with PEG (Ratanapongleka, 2010). The most common polymer-salt system used is PEG and phosphate salt (generally sodium or potassium phosphate) system due to low cost. Some examples of polymer-salts ATPSs usage are given in Table 2.

Table 1. Examples of polymer-polymer ATPSs

Bioproduct	Source	ATPS composition	Reference
PCV2 Cap protein	porcine circovirus type 2 Cap protein fermentation broth	$P_{ADB4.99}/P_{MDM7.08}$ (4/6% w/w)	He et al., 2018
transglutaminase	crude transglutaminase	$P_{ADB4.91}/P_{ADB4.06}$ (3/2% w/w)	Dong et al., 2018
mandelic acid enantiomers	mandelic acid chiral mixture	poly(MAH-β-CD-co-NIPAAm)/dextran (10/10% w/w)	Tan et al., 2017
proteins	protein mixture	PEG 8000/dextran (12.4/6.1% w/w)	Madeira et al., 2008
clavulanic acid	*Streptomyces clavuligerus* fermentation broth	PEG 4000/NaPA 8000 (10/20% w/w)	Pereira et al., 2012

*ADB polymer synthesized using acrylic acid (AA), N,N-dimethylaminoethyl methacrylate (DMAEMA) and butyl methacrylate (BMA).

**MDM polymer synthesized using methacrylic acid (MAA), N,N-dimethylaminoethyl methacrylate (DMAEMA) and methyl methacrylate (MMA).

***NaPA sodium polyacrylate.

Table 2. Examples of polymer-salts ATPSs

Bioproduct	Source	ATPS composition	Reference
crude protein	aqua waste extract	PEG 600/$(NH_4)_2SO_4$ (20/20% w/w)	Baskaran et al., 2018
cerein 8	*B. cereus* 8A culture supernatant	PEG 6000/$(NH_4)_2SO_4$ (20/20% w/w)	Lappe et al., 2012
polygalacturonase	*Aspergillus aculeatus* URM4953	PEG/citrate (19/23% w/w)	de Carvalho Silva et al., 2018
pancreatic trypsin	pancreas	PEG 3350/NaCit (40/25% w/w)	Perez et al., 2015
sulfated polysaccharides	crude natural mixture	PEG 1000 /$(NH_4)_2SO_4$ (25.2/18.76% w/w)	Du et al., 2018

TWO PHASE SYSTEMS FORMATION AND PARTITIONING

ATPSs are prepared by dissolving polymers in aqueous electrolyte and by intensive shaking of the mixture. After that, the systems are allowed to settle for some hours during which phase separation occurs. In case of PEG-dextran ATPS, the bottom phase is dextran phase, while the top phase is PEG phase. When using the ATPS for separation, the sample is contained in one of the phases (Figure 1).

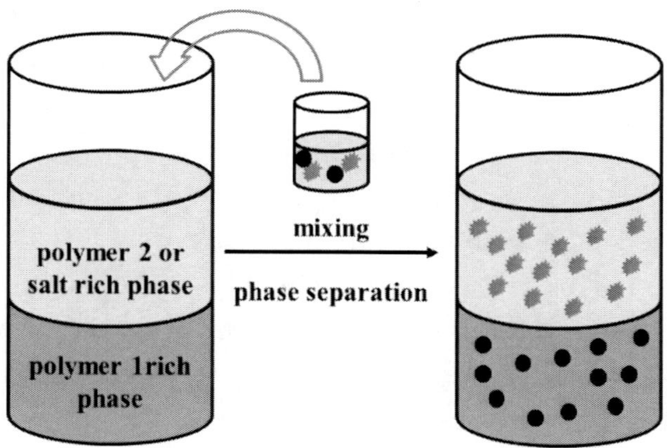

Figure 1. Phase separation based on ATPS (inspired by Wijethunga and Moon, 2015).

Partitioning in the ATPSs depends on electrostatic, hydrophobic and steric hindrance interactions. Partitioning in the ATPSs may be used as a bioanalytical tool for characterization of protein surface properties, changes in protein structure, conformation, etc. Understanding the mechanism of solute partitioning in the ATPSs is important for both separation and bioanalytical applications. Partition in the ATPSs can be described by Boltzman equation, if the particle diffuses form one phase to another:

$$K = \frac{n_1}{n_2} = \frac{c_1}{c_2} = e^{-\frac{\Delta E}{k \cdot T}}$$

where K presents a partition coefficient, n_1 and n_2 are the number of the particles in the phase 1 and phase 2, c_1 and c_2 are the concentrations of the particle in phase 1 and 2, ΔE is energy needed to move the particle to another state, k Boltzman's constant and T is the temperature (Shin et al., 2015).

According to Asenjo and Andrews (2012) the factors that influence partitioning in ATPSs are: (i) molecular weight/size of polymers, (ii) concentration of polymer, (iii) ionic strength of the salt, (iv) pH, and (v) additional salts used. With an increase of the PEG's molecular weight (PEG is widely used, nontoxic and relatively low cost polymer), density also increases (Wysoczanska and Macedo, 2016). According to Wu et al., (1998) the higher the molecular weight of the polymers the lower the concentration needed for the formation of two phases, and the larger the difference in molecular weights between the polymers, the more asymmetrical is the curve of the phase diagram. To form two phases, the polymer concentration is usually varied in range from 8% to 18% (w/w). Increasing the polymer concentration will result in the increase of density and phase viscosity (Ketnawa et al., 2017). The pH of the systems has the influence on the partition coefficients because it changes the charge of the solute or it changes the charge of the molecules (Raja et al., 2011). It is also important to mention that in systems without salt addition, two-phase formation occurs only at very high polymer concentration, but by addition of salt, phase separation can occur at significantly lower polymer concentrations (3-5% w/w of each polymer) (Johansson et al., 2011).

MICROSTRUCTURED DEVICES

Micro total analysis systems (μ-TAS) or "lab-on-a-chip" have been intensively used over the past few years. To develop the efficient μ-TAS, it is important to analyze each of the structure units. Microstructured devices or microfluidic devices are processing units characterized by continuous flow through regular domains with characteristic dimensions in the sub-millimeter range (Suryawanshi et al., 2018). Fluidic channels and chambers that are tens to hundreds of micrometers wide are the defining characteristics of microfluidic devices (Streets and Huang, 2014). Small volumes have numerous advantages comparing to classical devices: (i) small amount of chemicals needed, (ii) effective heat and mass transfer, (iii) precise control of process conditions (iii) high surface to volume ratio (Mark et al., 2010).

Microstructured devices are fabricated in a range of materials from ceramics, over the polymers, stainless steel and silicone to glass. Based on the construction of the basic unit, microstructured devices can be divided on chip-type microreactors and microcapillary devices (Yao et al., 2009). According to Yao et al., (2009) microfluidic devices were primarily used in high throughput screening in microanalytical chemistry, biological analysis of cells and proteins, reaction kinetics and mechanism studies.

In order to define the optimal process conditions in microstructured devices, flow pattern or flow profile is first to be analyzed. When working with two liquids in the microstructured device, as the aqueous two phase systems are, two flow patterns occur; parallel flow and slug flow. According to Dessimoz et al., (2008) the main problem in the control of flow pattern is the flow dependence on the experimental parameters like (i) the linear velocity, (ii) the ratio of phases, (iii) the fluid properties and (iv) the microchannel geometry. For the two-phase laminar flow it is crucial to form a stable liquid-liquid interface in microfluidic channel. In case of ATPSs stable parallel laminar flow is formed due to the low interfacial tension of the phases.

In parallel laminar two-phase flow in microstructured device, mass transfer is governed by molecular diffusion at Reynolds number less than 1. Fast mass transfer is one of the key characteristics of the microstructured devices. Due to the small dimensions, microstructured devices are characterized by short diffusion path that allows a rapid mass transfer and a uniform solute concentration within the flow domain (Hardt and Hahn, 2012).

APPLICATION OF ATPSS IN MICROSTRUCTURED DEVICES

Proteins and Bioactive Molecules Extraction, Purification and Analysis

Proteins separation technology using microfluidic devices was developed for more rapid and effective analysis of target protein. Huh et al., (2010) presented a new microfluidic protein separation technology for obtaining highly purified proteins. The proposed microfluidic system was used for the enzymes purification with ATPSs. Selected ATPS ensured stable three-phase stream through microchannel length. Novak et al., (2012) presented continuous extraction of proteins in different aqueous two-phase systems. Ionic liquid based ATPSs and PEG/phosphate ATPSs were compared.

Results showed that ionic liquid based ATPSs have several advantages comparing to PEG/phosphate ATPSs. Proposed system ensured bovine serum albumin extraction factor of 53% for retention time of 175 s. Polyethylene glycol (PEG)/phosphate two-phase system was used for continuous extraction of α-amylase in microfluidic chip (Novak et al., 2015). Proposed microextractor ensured two orders of magnitude shorter time for achieving the steady state comparing to batch extraction. Hu et al., (2011) described the usage of a PEG/detergent μ-ATPS system for the purification of the membrane proteins from crude cell extract and membrane proteins from HeLa cell extracts. The proposed PEG/detergent two-phase system portioning allowed successful removal of soluble proteins.

Aqueous two-phase systems in microfluidic devices have also shown potential for downstream processing of monoclonal antibodies. As described by Silva et al., (2012), the application of PEG/phosphate buffer with NaCl was investigated for the partition of immunoglobulin G in PDMS microfluidic device. Partition of immunoglobulin G from salt rich phase to PEG rich phase was measured by fluorescence microscopy due the stable interphase. Comparing to conventional extraction process, reduced operational time was noticed. Microfluidic devices have also great potential as the technology for fast detection of contaminants. Example for that is given by Soares et al., (2017) where novel sample preparation methodology based on PEG/sodium citrate ATPSs in microfluidic device is described. Proposed methodology ensured single-step extraction and concentration of three mycotoxins (aflatoxin, ochratoxin and deoxynivalenol) within 20 minutes. In paper by Bras et al., (2017) a microfluidic chip for simultaneously screening eight independent ATPSs is presented. The microfluidic toolbox was used to demonstrate the potential of LYTAG fusion proteins used for the affinity tags to optimize the partitioning of antibodies.

Results indicated that approximately 3.7-fold higher value of partition coefficient was obtained when using LYTAG fusion proteins comparing to the process without affinity molecule. Meng at al. (2018) described the continuous-flow process based on an ATPSs parallel and laminar flow in microfluidic device used for enzymatic catalysis. Due to the numerous advantages of microreactors, enzymatic reaction in ATPSs microfluidic devices is 500-fold higher comparing to conventional ATPSs in beaker under mixing.

Aqueous two-phase systems in microfluidic devices have also found their application in extractions of bioactive molecules. Šalić et al., (2011) used PEG 6000/$(NH_4)_2SO_4$ aqueous two-phase system for extraction of polyphenols for white and red wine. Analyzed geometry of the microextractor ensured 60-fold shorter residence time for achieving the same partition coefficient and extraction efficiency comparing to macroextractor.

Vobecká et al., (2018) described the application of the ATPSs for the production of 6-aminopenicillanic acid (6-APA) catalyzed by penicillin acylase, followed by the extractive separation of 6-APA from the reaction mixture. After optimizing the ATPS system composition, extraction experiments in microfluidic device showed that the transport of 6-APA, formed in the salt-rich phase to the corresponding PEG phase, could occur within 30 seconds.

CELL EXTRACTION AND SEPARATION

Microfluidic devices with ATPSs have also been used for the separation of blood components. SooHoo and Walker (2007) presented the separation of leukocyte from whole blood. The separation of the blood cells was based on the high specificity of lysing solution that was added into PEG stream. The lysing solution diffuses across the interface and lyses the erythrocytes. The leukocytes are held in place by the surface tension of the two-phase system while the remains of the lysed erythrocytes, other blood materials and the lysing solution are allowed to diffuse away and then diverted to another channel. The same authors (SooHoo and Walker, 2009) described the separation of leukocytes from a whole blood sample using microfluidic aqueous two-phase system PEG/dextran. Blood cells were partitioned based on their differential affinity for the streams in laminar flow regime. The used microfluidic device ensured up to a 9.13-fold increase in concentration of leukocytes comparing to classical batch extraction. Another example of the efficient use of the aqueous two-phase systems for cell separation in microfluidic device is given by Tsukamoto et al., (2009). Aqueous two-phase (PEG/dextran) laminar flow in a microfluidic chip was used to isolate leukocyte and erythrocyte cells from whole blood cells. When whole blood cells were inserted into the microfluidic chip, leukocytes could be separated from erythrocytes because erythrocytes moved to the dextran layer while leukocytes remained outside of this layer in the microfluidic system. Based on that results it can be concluded that whole blood cell separation can effectively be integrated into μ-TAS designed for cell-based clinical, forensic, and environmental analyzes.

Beside blood cell separation, ATPSs in microfluidic devices are also being used for the separation of the microorganism's cell. Periyannan et al., (2011) described the usage of PEG/dextran ATPS for partitioning of two strains of *E. coli*. Results have shown that two strains show different affinity for the phase, one strain was concentrated in the PEG phase, while the other in the dextrane phase. Nam et al., (2005) described the application of the microfluidic devices for the separation of the animal cell.

Live and dead CHO-K1 (Chinese Hamster Ovary) cells were fractionated by continuous-flow extraction in microfluidic device using aqueous two-phase system. Live cells were distributed to PEG rich phase, while the dead cells were found at the interface of the two polymer solutions.

PREPARING ATPSs IN MICROFLUIDIC DEVICES

Beside intensive researches in application of ATPSs for extraction and separation, lately there has been growing interest in preparing and characterizing ATPSs in microfluidic devices. Characterization of ATPSs by using lab-on-a-chip technology in combination with robotic liquid handling station is reported by Amrhein et al., (2014). According to Amrhein et al., (2014) by using the methodology described in their papers, two complete ATPSs could be characterized within 24 h, including four runs per ATPS for binodal curve determination (less than 45 min/run), and tie line determination (less than 45 min/run for ATPS preparation and 8 h for density determination), which can be performed fully automated over night without requiring man power.

Wijathunga and Moon (2015) presented the potential of preparing ATPSs using EWOD (electrowetting on dielectric) digital microfluidics. The experiments were performed by dispersing the equal volume of PEG and DEX solutions and getting them into contact to form an ATPS. The merged liquids were forced to mix by moving them along electrodes where different droplets were formed. The obtained results demonstrated that the quality of the ATPSs formed on-chip is very close to that of ATPSs formed at macroscale. The microfluidic system that generates water-in-water, aqueous two-phase system (ATPS) droplets was described by Moon et al., (2016). ATPS droplets formation was achieved by applying weak hydrostatic pressure with liquid-filled pipette tips as fluid columns at the inlets to introduce low speed flows to the flow-focusing junction. The proposed system shows the potential application for encapsulating cells in water-water emulsions by encapsulating microparticles and cells.

Mastiani et al., (2017) tested the effect of the different micro device geometries on properties and size of formed ATPS droplets. Results showed that microreactor inlet junction angle has significant effect on the droplet size; flow regime mapping for two different droplet generators at 30° and 90° were also presented. A method for characterization of slug flow of two immiscible aqueous phases in flow channel was proposed by Polezhaev et al., (2018).

Electrochemical impedance spectroscopy based on the measuring impedance response of the two-phase system was described. Results indicate that developed electrochemical impedance spectroscopy method is an original approach for estimation of wall fill thickness useful in control of various droplet based microfluidic devices. Simple approach for the production of the uniformed ATPS droplets facilitated by oil droplet choppers in microfluidics is presented by Zhou et al., (2017). The authors described the synchronized formation of high interfacial tension oil-in-water and low-interfacial-tension water-in water droplets where the ATPS interface is distorted by oil droplets and decays into water-in water droplets. The authors declare that the proposed methodology would boost emulsion-based biological application such as cell encapsulation, biomolecule delivery, bioreactors and biomaterials synthesis with ATPS droplets.

CONCLUSION

Compared with conventional liquid–liquid extraction using aqueous/organic extraction media, aqueous two-phase system is known to provide relatively easy mass transfer and a gentle environment for biological separation applications. Recent interest in microscale ATPSs is aimed to study biomolecule extraction and separation, cell separations and preparing ATPS droplets in microfluidic device.

REFERENCES

Almeida, M. R., Passos, H., Pereira, M. M., Lima, A. S., Coutinho, J. A. P. & Freire, M. G. (2014). Ionic liquids as additives to enhance the extraction of antioxidants in aqueous two-phase systems. *Separation and Purification Technology, 128*, 1-10.

Álvarez, M. S, Moscoso, F., Rodríguez, A., Sanromán, M. A. & Deive, F. J. (2012). Triton X surfactants to form aqueous biphasic systems: Experiment and correlation. *Journal of Chemical Thermodynamics, 54*, 385-392.

Amrhein, S., Schwab, M.-L., Hoffman, M. & Hubbuch, J. (2014). Characterization of aqueous two-phase systems by combining lab-on-a-chip technology with robotic liquid handling stations. *Journal of Chromatography A, 1367*, 68-77.

Asenjo, J. A. & Andrews, B. (2012). Aqueous two-phase systems for protein separation: Phase separation and applications. *Journal of Chromatography A*, 1238, 1-10.

Baskaran, D., Chinnappan, K., Manivasagan, R. & Mahadevan Kumar. D. (2018). Partitioning of crude protein from aqua waste using PEG 600-inorganic salt aqueous two-phase systems. *Chemical Data Collections, 15*, 143-153.

Bras, E. J., Soares, R. R. G., Azevedo, A. M., Fernandez, P., Arevalo-Rodriguez, M., Chu, V., Conde, J. P. & Aires-Barros, M. R. (2017). A multiplexed microfluidic toolbox for the rapid optimization of affinity-driven partition in aqueous two-phase systems. *Journal of Chromatography A, 1515*, 252-259.

de Carvalho Silva, J., Lopes de Franca, P. R. & Porto, T. S. (2018). Optimized extraction of polygalacturonase from *Aspergillus aculeatus* URM4953 by aqueous two-phase systems PEG/Citrate. *Journal of Molecular Liquids, 263*, 81-88.

Dessimoz, A.-L., Cavin, L., Renken, A. & Kiwi, L. (2008). Liquid–liquid two-phase flow patterns and mass transfer characteristics in rectangular glass microreactors. *Chemical Engineering Science, 63*, 4035-4044.

Dong, L., Wan, J. & Cao, X. (2018). Separation of transglutaminase using aqueous two-phase systems composed of two pH-response polymers. *Journal of Chromatography A, 1555*, 106-112.

Du, L.-P., Cheong, K.-L. & Liu, Y. (2018). Optimization of an aqueous two-phase extraction method for the selective separation of sulfated polysaccharides from a crude natural mixture. *Separation and Purification Technology, 202*, 290-298.

Goja, A. M., Yang, H., Cui, M. & Li, C. (2013). Aqueous two-phase extraction advances for bioseparation. *Journal of Bioprocessing and Biotechniques, 4*, 1-6.

Hardt, S. & Hahn, T. (2012). Microfluidics with aqueous two-phase systems. *Lab on Chip, 12*, 434-443.

He, J., Wan, J., Yang, T., Cao, X. & Yang, L. (2018). Recyclable aqueous two-phase system based on two pH-responsive copolymers and its application to porcine circovirus type 2 Cap protein purification. *Journal of Chromatography A, 1555*, 113-123.

Hu, R., Feng, X., Chen, P., Fu, M., Chen, H., Guo, L. & Liu, B.-F. (2011). Rapid, highly efficient extraction and purification of membrane proteins using a microfluidic continuous-flow based aqueous two-phase system. *Journal of Chromatography A, 1218*, 171-177.

Huh, Y. S., Jeong, C.-M., Chang, H. N., Lee, S. Y., Hong, W. H. & Park, T. J. (2010). Rapid separation of bacteriorhodopsin using a laminar-flow extraction system in a microfluidic device. *Biomicrofluidics, 4*, 1-11.

Johanson H.-O., Feitosa, E. & Pessoa Jounior, A. (2011). Phase diagrams of the aqueous two-phase systems of poly(ethylene glycol)/sodium polyacrylate/salts. *Polymers, 3*, 587-601.

Ketnawa, S., Rungraeng, N. & Rawdkuen, S. (2017). Phase partitioning for enzyme separation: An overview and recent applications. *International Food Research Journal, 24*, 1-24.

Lappe, R., Sant'Anna, V. & Brandelli, A. (2012). Extraction of the antimicrobial peptide cerein 8A by aqueous two-phase systems and aqueous two-phase micellar systems. *Natural Product Research, 26*, 2259–2265.

Liu, Y., Lipowsky, R. & Dimova, R. (2012). Concentration dependence of the interfacial tension for aqueous two-phase polymer solutions of dextran and polyethylene glycol. *Langmuir, 28*, 3831–3839.

Lladosa, E., Silverio, S. C., Rodriguez, O., Teixeira, J. A. & Macedo, E. A. (2012). (Liquid + liquid) equilibria of polymer-salt aqueous two-phase systems for laccase partitioning: UCON 50-HB-5100 with potassium citrate and (sodium or potassium) formate at 23 °C. *Journal of Chemical Thermodynamics, 55*, 166-171.

Madeira, P. P., Teixeira, J. A., Macedo, E. A., Mikheeva, L. M. & Zaslavsky, B. Y. (2008). "On the Collander equation": Protein partitioning in polymer/polymer aqueous two-phase systems. *Journal of Chromatography A, 1190*, 39-43.

Mark, D., Haeberle, S., Roth, G., von Stetten, F. & Zengerle, R. (2010). Microfluidic lab-on-a-chip platforms: Requirements, characteristics and applications. *Chemical Society Reviews, 39*, 1153-1182.

Mastiani, M., Seo, S., Jimenez, S. M., Petrozzi, N. & Kim, M. M. (2017). Flow regime mapping of aqueous two-phase system droplets in flow-focusing geometries. *Colloids and Surfaces A: Physicochemical and Engineering Aspects, 531*, 111-120.

Meng, S.-X., Xue, L.-H., Xie, C.-Y., Bai, R.-X., Yang, X., Qiu, Z.-P., Guo, T., Wang, Y.-L. & Meng, T. (2018). Enhanced enzymatic reaction by aqueous two-phase systems using parallel-laminar flow in a double Y-branched microfluidic device. *Chemical Engineering Journal, 335*, 392-400.

Moon, B.-U., Abbasi, N., Jones, S. G., Hwang, D. K. & Tsai, S. S. H. (2016) Water-in-water droplets by passive microfluidic flow focusing. *Analytical Chemistry, 88*, 3982-3989.

Nam, K. H., Chang, W. J., Hong, H., Lim, S. M., Kim, D. I. & Koo, Y. M. (2005). Continuous-flow fractionation of animal cells in microfluidic device using aqueous two-phase extraction. *Biomedical Microdevices, 7*, 189-95.

Novak, U., Lakner, M., Plazl, I. & Žnidaršič-Plazl, P. (2015). Experimental studies and modelling of α-amylase aqueous two-phase extraction within a microfluidic devices. *Microfluidics and Nanofluidics, 19,* 75-83.

Novak, U., Pohar, A., Plazl, I. & Žnidaršič-Plazl, P. (2012). Ionic liquid-based aqueous two-phase extraction within a microchannel system. *Separation and Purification Technology, 97,* 172-178.

Pereira, J. F. B., Santos, V. C., Johansson, H.-O., Teixeira, J. A. C. & Pessoa Jr., A. (2012). A stable liquid–liquid extraction system for clavulanic acid using polymer-based aqueous two-phase systems. *Separation and Purification Technology, 98,* 441-450.

Perez, R. L., Loureiro, D. B., Nerli, B. B. & Tubio, G. (2015). Optimization of pancreatic trypsin extraction in PEG/citrate aqueous two-phase systems. *Protein Expression and Purification, 106,* 66-71.

Periyannan, P. K., Ramachandraiah, H., Hansson, J., Ardabili, S., Veide, A. & Russom, A. (2011). Development of microfluidic aqueous two-phase system for continuous partitioning of *E. coli* strains. *15th International Conference on Miniaturized Systems for Chemistry and Life Sciences,* October 2-6, 2011, Seattle, Washington, USA, 1329-1331.

Polezhaev, P., Slouka, Z., Lindner, J. & Přibly, M. (2018). Characterization of slug flow of two aqueous phases by electrochemical impedance spectroscopy in a fluidic chip. *Microelectronic Engineering, 194,* 89-95.

Rahimpour, F. & Baharvand, A. R. (2009). Phase equilibrium in aqueous two-phase systems containing poly(propylene glycol) and sodium citrate at different pH. *World Academy of Science, Engineering and Technology, 35,* 150-154.

Raja, S., Murty, V. R., Thivaharan, V., Rajasekar, V. &Ramesh, V. (2011). Aqueous two-phase systems for the recovery of biomolecules – A review. *Science and Technology, 1,* 7-16.

Ratanapongleka, K. (2010). Recovery of biological products in aqueous two-phase systems. *International Journal of Chemical Engineering and Applications, 1,* 191-199.

Rodrigues de Lemos, L., Santos, I. J. B., Rodrigues, G. D., Dias Ferreira, G. M., Mendes da Silva, L. H., Hespanhol da Silva, M. C. & Maduro de Carvalho, R. M. (2010). Phase compositions of aqueous two-phase systems formed by L35 and salts at different Temperatures. *Journal of Chemical Engineering Data, 55*, 1193-1199.

Shin, H., Han, C., Labuz, J. M., Kim, J., Kim, J., Cho, S., Gho, Y. S., Takayama, S. & Park, J. (2015). High-yield isolation of extracellular vesicles using aqueous two-phase system. *Scientific Reports, 5*, 1-11.

Silva, D. F. C., Azevedo, A. M., Fernandes, P., Chu, V., Conde, J. P. & Aires-Barros, M.R. (2012). Design of a microfluidic platform for monoclonal antibody extraction using an aqueous two-phase system. *Journal of Chromatography A, 1249*, 1-7.

Silverio, S. C., Moreira, S., Milagres, A. M. F., Macedo, E. A., Teixeira, J.A. & Mussatto, S. I. (2012). Interference of some aqueous two-phase system phase-forming components in protein determination by the Bradford method. *Analytical Biochemistry, 421*, 719-724.

Silvério, S. C., Rodríguez, O., Tavares, A. P. M., Teixeira, J. A. & Macedo, E. A. (2013). Laccase recovery with aqueous two-phase systems: Enzyme partitioning and stability. *Journal of Molecular Catalysis B: Enzymatic, 87*, 37-43.

Soares, R. R. G., Azevedo, A. M., Fernandes, P., Chu, V., Conde, J. P. & Aires-Barros, M. R. (2017). A simple method for point-of-need extraction, concentration and rapid multi-mycotoxin immunodetection in feeds using aqueous two-phase systems. *Journal of Chromatography A, 1511*, 15-24.

SooHoo, J. R. & Walker, G. M. (2007). Microfluidic liquid filters for leukocyte isolation. Conference proceedings. *Annual International Conference of the IEEE Engineering in Medicine and Biology Society, 2007*, 6319-6322.

SooHoo, J. R. & Walker, G. M. (2009). Microfluidic aqueous two-phase system for leukocyte concentration from whole blood. *Biomedicine Microdevices, 11*, 323–329.

Streets, A. M. & Huang, Y. (2014). Microfluidics for biological measurements with single-molecule resolution. *Current Opinion in Biotechnology, 25*, 69-77.

Suryawanshi, P. L., Gumfekar, S. P., Bhanvase, B. A., Sonawane, S. & Pimplapure, M. P. (2018). A review on microreactors: Reactor fabrication, design, and cutting-edge applications. *Chemical Engineering Science.* DOI: 10.1016/j.ces.2018.03.026.

Sutherland, I., Hewitson, P., Siebers, R., van den Heuvel, R., Arbenz, L., Kinkel, J. & Fisher, D. (2011). Scale-up of protein purifications using aqueous two-phase systems: Comparing multilayer toroidal coil chromatography with centrifugal partition chromatography. *Journal of Chromatography A, 1218*, 5527-553.

Šalić, A., Tušek, A., Fabek, D., Rukavina, I. & Zelić, B. (2011). Aqueous two phase extraction of polyphenols using a microchannel system – Process optimization and intensification. *Food Technology and Biotechnology, 49,* 495-501.

Tan, Z., Li, F., Zhao, C., Teng, Y. & Liu, Y. (2017). Chiral separation of mandelic acid enantiomers using an aqueous two-phase system based on a thermo-sensitive polymer and dextran. *Separation and Purification Technology, 172,* 382-387.

Teixeira, A. G., Agarwal, R., Ko, K. R., Grant-Burt, J., Leung, B. M. & Frampton, J. P. (2017). Emerging biotechnology applications of aqueous two-phase systems. *Advanced Healthcare Materials, 2017,* 1-19.

Tsukamoto, M., Taira, S., Yamamura, S., Morita, Y., Nagatani, N., Takamura, Y. & Tamiya, E. (2009). Cell separation by an aqueous two-phase system in a microfluidic device. *Analyst- Royal Society of Chemistry, 134,* 1994-1998.

Vobecká, L., Romanov, A. & Slouka, Z. (2018). Optimization of aqueous two-phase systems for the production of 6-aminopenicillanic acid in integrated microfluidic reactors-separators. *New Biotechnology,* DOI: 10.1016/j.nbt.2018.03.005.

Wijethunga, P. A. L. & Moon, H. (2015). On-chip aqueous two-phase system (ATPS) formation, consequential self-mixing, and their influence on drop-to-drop aqueous two-phase extraction kinetics. *Journal of Micromechanics and Microengineering, 25,* 094002-094013.

Wu, Y. T., Lin, D. Q. & Zhu, Z. Q. (1998). Thermodynamics of aqueous two-phase systems - the effect of polymer molecular weight on liquid-liquid equilibrium phase diagrams by the modified NRTL model. *Fluid Phase Equilibria, 147*, 25-43.

Wysoczanska, K., Macedo, E. A. (2016). Influence of the molecular weight of PEG on the polymer/salt phase diagrams of aqueous two-phase systems. *Journal of Chemical Engineering Data, 61*, 4229-4235.

Xu, Y., Souza, M. A., Pontes, M. Z. R., Vitolo, M. & Pessoa Junior, A. (2003). Liquid-liquid extraction of enzymes by affinity aqueous two-phase systems. *Brazilian Archives of Biology and Technology, 46*, 741-750.

Yao, X., Zhang, Y., Du, L., Liu, J. & Yao, J. (2009). Review of the applications of microreactors. *Renewable and Sustainable Energy Reviews, 47*, 519-539.

Zhou, C., Zhu, P., Tian, Y., Tang, X., Shi, R. & Wang, L. (2017). Microfluidic generation of aqueous two-phase-system (ATPS) droplets by oil-droplet choppers. *Lab on Chip, 17*, 3310-3317.

INDEX

A

acid, 13, 35, 40, 46, 48, 49, 52, 92, 97, 104, 106

affinity of the proteins, 31

ammonium, 24, 30, 35, 36, 40, 48, 53, 87, 90, 91

aqueous biphasic systems, vii, ix, 16, 19, 22, 42, 43, 44, 46, 47, 48, 50, 52, 53, 84, 85, 86, 87, 88, 90, 101

aqueous mixture of cells, ix, 21

aqueous PEG-based two-phase systems, vii, x, 55, 61, 65, 66, 68, 70, 72, 73, 74, 75, 76, 77, 78, 79, 80, 81, 83

aqueous solutions, 62

aqueous two-phase system, vii, x, 1, 3, 4, 5, 7, 8, 9, 10, 11, 12, 13, 14, 15, 16, 17, 18, 19, 22, 24, 25, 30, 35, 36, 40, 42, 43, 44, 45, 46, 47, 48, 49, 50, 51, 52, 53, 56, 57, 58, 59, 61, 64, 65, 66, 68, 69, 70, 71, 72, 73, 74, 76, 77, 78, 82, 83, 84, 85, 86, 87, 89, 90, 96, 97, 98, 99, 100, 101, 102, 103, 104, 105, 106, 107

ATPS partitioning, 31

B

bacteria, 12, 33

binodal curve, 7, 26, 27, 28, 99

biocompatibility, ix, 22, 31

bioconversion, 34

biodegradation, 47

biodiesel, 45, 47

biological activity, viii, ix, 2, 3, 4, 22

biological fluids, 16

biological samples, 14

biomaterials, 100

biomolecule partition study, 26

biomolecules, viii, ix, x, 1, 4, 5, 8, 13, 14, 18, 22, 24, 25, 31, 32, 38, 39, 41, 49, 50, 89, 91, 104

bioremediation, 29

bioseparation, 102

biosynthesis, 37, 48

biotechnological applications, 33

biotechnological processes, ix, 21, 39

biotechnology, 13, 15, 17, 22, 29, 30, 42, 43, 44, 45, 47, 48, 49, 50, 52, 89, 91, 105, 106

blood, 19, 98, 105

butyl methacrylate, 92

R

S

T

V

W